Palgrave Studies in Media and Environmental Communication

Series Editors
Anders Hansen
Department of Media and Communication
University of Leicester
Leicester, UK

Steve Depoe
McMicken College of Arts & Sciences
University of Cincinnati
Cincinnati, OH, USA

Drawing on both leading and emerging scholars of environmental communication, the Palgrave Studies in Media and Environmental Communication Series features books on the key roles of media and communication processes in relation to a broad range of global as well as national/local environmental issues, crises and disasters. Characteristic of the cross-disciplinary nature of environmental communication, the books showcase a broad variety of theories, methods and perspectives for the study of media and communication processes regarding the environment. Common to these is the endeavour to describe, analyse, understand and explain the centrality of media and communication processes to public and political action on the environment.

More information about this series at
http://www.palgrave.com/gp/series/14612

Roy Bendor

Interactive Media
for Sustainability

palgrave
macmillan

Roy Bendor
Delft University of Technology
Delft, The Netherlands

Palgrave Studies in Media and Environmental Communication
ISBN 978-3-030-09954-1   ISBN 978-3-319-70383-1   (eBook)
https://doi.org/10.1007/978-3-319-70383-1

Cover illustration: Vladimir Godnik

Printed on acid-free paper

This Palgrave Macmillan imprint is published by the registered company Springer Nature Switzerland AG
The registered company address is: Gewerbestrasse 11, 6330 Cham, Switzerland

To Tracy, Noam, and Zohar

ACKNOWLEDGMENTS

Writing this book was a wonderful, even if challenging, journey. I could have not completed it without the encouragement and support of my colleagues, friends, and family.

Andrew Feenberg's clear vision and inspiring scholarship were incredible resources to draw from. Shane Gunster lent me his critical ear and kind, calm advice when I needed it the most. And John Robinson's generosity and unorthodox thinking were essential to the project's conceptualization and completion.

The ideas developed here took shape through discussions with colleagues at Simon Fraser University's School of Communication (the ACT Lab in particular), the University of British Columbia's Centre for Interactive Research on Sustainability, the idStudioLab at Delft University of Technology, and the Disruptive Imaginings collaborative (led by the indefatigable Vanessa Timmer). Conversations with collaborators in the *Greenest City Conversations*, and *Sustainability in an Imaginary World* projects (David Maggs and Steve Williams in particular) were especially valuable. Boudewijn Boon and Marco Rozendaal made insightful comments on draft chapters. Darryl Cressman, Ted Hamilton, Rob Prey, Neal Thomas, and Joost Vervoort were generous with their time, and critical and supportive with their advice.

My research has been supported by the Social Sciences and Humanities Research Council of Canada (SSHRC). My colleagues at Delft University of Technology's Department of Industrial Design (David Keyson, Ingrid

Mulder, and Joost Niermeijer in particular) showed understanding and compassion when they provided me with the conditions necessary to complete this book even as I transitioned to a new, demanding position.

Last, but certainly not least, through the ups and downs of the project, my family kept me grounded and reminded me why I started writing this book in the first place. Their unwavering love and support means the world to me.

CONTENTS

Acronyms and Abbreviations

3D	Three-Dimensional
AR	Augmented Reality
BC	British Columbia (province in Canada)
CBC	Canadian Broadcasting Corporation
CIRS	Centre for Interactive Research on Sustainability (at the University of British Columbia)
CO_2	Carbon Dioxide
DfBC	Design for Behavior Change
EMS	Electronic Muscle Stimulation
GHG	Greenhouse Gases
GPS	Global Positioning System
GUI	Graphic User Interface
HCI	Human Computer Interaction
ICT	Information and Communication Technology
ICT4S	Information and Communication Technology for Sustainability
IPCC	Intergovernmental Panel on Climate Change
IT	Information Technology
MIT	Massachusetts Institute of Technology
MP3	Media Player 3 (file format)
MQ	MetroQuest (tool and company)
MQ-V	Vancouver version of the MetroQuest tool
NIMBY	Not in My Back Yard
SHCI	Sustainable Human Computer Interaction
SLE	Significant Life Experiences

TCP/IP   Transmission Control Protocol/Internet Protocol
UN   United Nations
UNWCED   United Nations World Commission on Environment and
          Development
VCR   Video Cassette Recorder
VR   Virtual Reality

LIST OF IMAGES

CHAPTER 1

Mediation

CATCHING UP OUR THINKING WITH OUR LIVING

Fifty years since sustainability became a neologism for long-term societal durability, and thirty years since it was formalized with the publication of the Brundtland Commission's report *Our Common Future* (1987), it is safe to say that sustainability is not a fad. Despite early concerns and detractions,[1] the term has not only caught on, but can now be seen gracing a variety of government units, academic programs, business practices, professional networks, and organizations of all stripes and colors. From a concept that was meant to call attention to the grotesque absurdity and latent death-drive of growth economics, the embodiment of a "frontiers worldview," as Johan Rockström (2015) puts it, sustainability has become a rallying cry for a social movement whose goal is to reimagine, reconfigure, and remake (the new "three R's"!) modern society.

As sustainability gained scientific traction and cultural cachet, it became apparent that both its meaning and its prospects were entangled with technology. Is this merely poetic retribution for an original sin, the proverbial technological chickens coming home to roost? After all, sustainability appears to be our best hope for solving an ecological crisis that resulted from the material processes, outcomes, and mindset associated

---

[1] Bill McKibben, for instance, wrote in 1996 that sustainability is a "buzzless buzzword" that would never catch on in "mainstream society" (see Caradonna 2014, p. 2).

© The Author(s) 2018
R. Bendor, *Interactive Media for Sustainability,*
Palgrave Studies in Media and Environmental Communication,
https://doi.org/10.1007/978-3-319-70383-1_1

with modern technology. While it may be debatable whether the seeds of the Anthropocene, a term meant to draw attention to the extent to which human activity is influencing the planet's geological processes, were already sown with the emergence of Homo Faber, with the development of ancient agriculture over 10,000 years ago, with the global intermingling of species that took place at the beginning of the seventeenth century, or with the industrial acceleration that took place in the 1960s (Lewis and Maslin 2015), there is little doubt that a direct path leads from the mass development and use of fossil-fueled industrial manufacturing to global climate change (IPCC 2015). This path was paved by a particular way of seeing the world as a growth-oriented machine, an ontological perspective that itself co-emerged with modern science and technology (Dijksterhuis 1961; Kumar 1978; Leiss 1972; Merchant 1989). As José Ortega y Gasset (1941) suggests, "The history of human thinking may be regarded as a long series of observations made to discover what latent possibilities the world offers for the construction of machines.... man begins where technology begins" (pp. 116–117). However, the entanglement of sustainability and technology reflects more than the ecological legacy of the Industrial Revolution. The roots of the entanglement may lie in the past, but its crown faces the future. This is because the field of activity represented by sustainability will certainly demand even more technological innovation and deployment. As scientists and economists tell us, a thorough decarbonization of the world economy will require investment in renewable energy, energy-efficient "smart" infrastructure, new ways to grow and transport food, smarter land-use practices, and new processes of industrial production (Figueres et al. 2017)—all necessitating various degrees of technological design and production. In this sense, to paraphrase German philosopher Martin Heidegger (1977), technology is both the "danger" of unsustainability and the "saving power" for a more sustainable future.

Claims that technology and sustainability are entangled, or that "sustainable interactive technology" is a paradoxical construction (Hazas and Nathan 2018a), may be true, but do little to illustrate the ways in which such an entanglement takes place. The issue is further complicated by two important factors: first, it is not always clear what we are talking about when we talk about "sustainability" (more on that below), or what we are referring to by "technology." The latter, in particular, conjures an expansive field of materials, artifacts, practices, ways of thinking, and ways of being—both a more general "activity form" and a set of particular technologies (McGinn 1990). While theorizations of technology in more general terms inform the ideas presented here, this book concerns new interactive technologies, sometimes referred to as information technology (IT), information and communication technology (ICT), digital media, interactive media, or simply, new media (despite the slight differences in emphasis each of these terms signal, they will be used here interchangeably). New media are practically everywhere: they are in our homes and in our cars, in our pockets and on our wrists. We use them to inquire, locate, share, and keep in touch. They compel us to take note and they nudge us to act. They influence our wants, dreams, and desires—desires that are often evoked and stoked in ways that only the media themselves can satisfy. New media are a fixture of our social imaginaries, part and parcel of how we see ourselves, others, and the world. And as I will argue here, new media have become inseparable from what sustainability means to us and, no less important, what it may mean in the future.

A second complication arises from the breathtaking pace of technological change. As soon as we start to grasp the implications of a new technology, a newer one appears on the horizon, deeming analytical efforts to a perpetual game of tag. The gap between sociotechnical praxis and reflection has long occupied the thought of critical technology scholars from Marx to Marcuse, Heidegger to Ellul, and it is eloquently captured in Marshall McLuhan's remark that "We are always living a way ahead of our thinking."[2] McLuhan's penchant for aphorisms notwithstanding, his point is as true today as it was when he made it in 1965. If anything, the velocity of our technological trajectory, accelerated by the volume of time, energy, and resources invested in its growth, makes it almost impossible to anticipate the entire gamut of implications introduced by new technologies. Even those we consider technology leaders often fail in their prognostications. IBM's president Thomas Watson has famously predicted in 1943 that there was a "world market for maybe five computers" (Pogue 2012, Jan. 18), and Ken Olsen, founder of Digital Equipment Corporation, stated in 1977 that "There is no reason anyone would want a computer in their home" (Skarda 2011, Oct. 21). Venerable publications such as MIT's

---

[2] The comment was made in an interview given to the Canadian Broadcasting Corporation (CBC) program, Take-30. It was broadcasted in April 1965, and is available here: http://www.youtube.com/watch?v=NNhRCRAL6sY (last accessed Mar. 18, 2018).

Technology Review have not done much better (Funk 2017). Setting aside modern technology's uncanny ability to turn erudite predictions on their head, every analysis of technology is firmly located in the time and place of its origin. Much like its subject, technological critique is situated and contingent, and what follows on these pages is no exception.

As strong as our anxieties over intellectual obsolescence may be, the pace of technological innovation should not deter but invite critical reflection. As we stand on the threshold of what has been dubbed the "fourth industrial revolution" (Schwab 2017), the future appears to be bursting with new technological possibilities and, doubtlessly, threats. Artificially intelligent agents and smart, interconnected objects, cancer-fighting nanobots, and autonomous vehicles, all promise new ways for improving the human condition while reducing humanity's ecological toll. But if we are about to inhabit a world in which the boundaries between the physical, the digital, and the biological blur beyond recognition, as Schwab (2017) forecasts, now would be an opportune moment to reflect, assess, and reorient technological innovation. In other words, if our thinking is to ever catch up with our living, we need to be able to identify both the fleeting and the enduring in technology, that which constantly changes as result of new materials, techniques, and applications, and that which orients technological design and use throughout. This is as relevant for sustainability as it is for any other field of human endeavor. Which social interests, values, and objectives will inform the design of future interactive technologies for sustainability? How can we prepare for the kind of moral and practical challenges new technologies will indubitably pose to those who use them to engage with sustainability? These are the questions that motivate the book's first aim: *to present a critical, conceptual framework for examining the contemporary design and use of interactive media for sustainability.*

WHERE HAVE WE BEEN?

Thankfully, I am not alone in trying to catch up our thinking with our living. Others have set their sights on the entanglement of technology and sustainability with promising, even if not entirely satisfactory, results. Two fields in particular have broached the topic: environmental communication and sustainable human-computer interaction (SHCI). The present work draws upon them while aiming to expand their purview.

After years of relative neglect,[3] environmental communicators have recently taken a keener interest in new media, seeing in them new opportunities for raising awareness of environmental issues, creating and disseminating relevant information, facilitating dialogue online, and providing nimble and cost-effective means to organize environmental activism (Moser 2010; O'Neill and Boykoff 2011; Schäfer 2012). Compared with "older" mass media, new media seem to open up unique possibilities for reaching both wider and narrower segments of the public while influencing the social contexts within which environmental messages are received. New media, especially social networks, are also valued for their capacity to extend virtually the environmental public sphere through a variety of dedicated and everyday spaces in which the public may express their opinions, debate the issues, come to common understandings, evoke a sense of collectivity or solidarity, and be motivated to act.

Given the field's social scientific character and affinity with studies of risk, rhetoric, journalism, and the mass media, it is not surprising that with a few exceptions (e.g., Büscher 2014; Fritsch and Brynskov 2011; Haider 2016; Starosielski and Walker 2016; Stephens et al. 2017), most analyses of new media for environmental communication focus on content and reach—how messages are constructed, framed, disseminated, interpreted, debated, and by whom.[4] These are all important aspects of new media's capacity to augment and even replace existing communicative channels, but what is often neglected in such accounts is the specificity of the media themselves. As discussed in more detail below, the specificity of new media

---

[3] New media or online communication are only scarcely mentioned in Moser and Dilling's path-breaking edited collection, *Creating a Climate for Change* (2007), and occupy a very small space in Hansen's *Environment, Media and Communication* (2010). They were oddly absent from an editorial dedicated to climate change communication research in *Applied Environmental Education & Communication* (Kelly 2012), did not fare much better in a special issue of *Environmental Communication* (vol. 8, issue 2, 2014) dedicated to the same topic, and are only mentioned sporadically in the most recent edition of *The Routledge Handbook of Environment and Communication* (Hansen and Cox 2015). There are reasons to believe the trend is changing with the more prominent space given to online communication in the latest (4th) edition of the influential *Environmental Communication and the Public Sphere* (Cox and Pezzullo 2015), and with the publication of a special issue of *Environmental Communication* (vol. 9, issue 2, 2015) dedicated to climate change and the internet.

[4] For a few recent examples, see Atanasova and Koteyko (2017); Anderson (2014); Cox and Pezzullo (2015); Katz-Kimchi and Manosevitch (2015); Kirilenko and Stepchenkova (2014); Pearce et al. (2014); Spartz et al. (2017).

is premised in their particular configuration of material, semiotic, and formal characteristics. Their interactivity, in particular, is crucial to the way they articulate function and meaning and thus mediate the world.[5] Without paying attention to the technological mediation, accounts of new media for environmental communication risk instrumentalism, a view of new media as neutral "'tools' standing ready to serve the purposes of their users … indifferent to the variety of ends" they could be used to achieve (Feenberg 2002, p. 5). Moser's (2016, p. 350) recent (albeit brief) discussion of "gaming and other interactive tools" as mere "communication aids" is a case in point.

Whereas environmental communicators have focused on the communicative or semiotic capacities of new media, designers and computer scientists have asked about the material consequences of new media's use. Much of that work, often referred to as "green computing," "sustainable ICT," or "ICT for Sustainability" (ICT4S), has focused on reducing the material impact of digital media—minimizing the environmental footprint implicated in the manufacture, use, and disassembly of computational devices, from the server farms that power the "cloud" to the mobile devices that have become our appendices.[6] Following Blevis's (2007) influential call to integrate sustainability "as a core semantics for interaction design" (p. 503), some in the SHCI community have shifted their attention from large-scale industrial processes onto individual consumer practices, and more recently onto the social, cultural, and political contexts within which individual consumption takes place (Dourish 2010; Fuad-Luke 2009; Knowles et al. 2014; Nardi and Ekebia 2018; Prost et al. 2014; Tomlinson 2010).[7] Current work in SHCI features granular accounts of new media's capacities, opening the "black box" of technology to evaluation and improvement by laying bare decisions about the scope, scale, and aims of design. However, as others have argued persuasively, SHCI has difficulties reaching beyond its all-too-familiar epistemologies and practices. The prevalence of "silo thinking" within the SHCI

---

[5] I am indebted here to Feenberg's notion of the "double aspects of technology," part of his instrumentalization theory (see Feenberg 1999, 2017).

[6] For a few recent examples, see Frick (2016); Hilty and Aebischer (2015); Issa et al. (2017); and the published proceedings of the International Conference on ICT for Sustainability (ICT4S), available at: http://ict4s.org/conference-proceedings (last accessed Mar. 10, 2018).

[7] For recent overviews of SHCI, see Silberman et al. (2014a, b) and Hazas and Nathan (2018b).

community is often expressed in calls for the community to engage with other, cognate fields of research and design, and to develop a more nuanced understanding of sustainability both in terms of its implications for design and in terms of its larger societal impacts (see, for instance, Busse et al. 2013; Clear et al. 2015; DiSalvo et al. 2010; Silberman et al. 2014a, 2014b). This, at times, manifests a palpable tension between the acknowledgment that sustainability can be interpreted in various, divergent ways, and the desire to have a singular, clear definition that could ground evaluative criteria.

Environmental communication's lack of attention to the specificities of technological mediation, and the SHCI community's need for a deeper, interdisciplinary engagement with the meaning of sustainability in theory and in practice, motivate the book's second aim: *to illustrate how, as a discursive field, the meaning of sustainability is increasingly pluralized as consequence of its mediation by interactive technologies.* Of course, questions pertaining to the meaning of sustainability have long been a topic of discussion among sustainability scholars and practitioners. Some have already pointed out that sustainability is an "essentially contested concept" (Connelly 2007; Jacobs 1999), one of those terms "like fairness, freedom, or liberty, for which there is some common sense for what they mean in the abstract, but which lack the same common sense of how to put them in play" (Ehrenfeld 2008, pp. 214–215). Others see this nebulosity as an invitation for misappropriation. Farley and Smith (2014, p. 3), for instance, argue that "the concept of sustainability has been overused, lacks clarity, and is vulnerable to co-option, thereby allowing sustainability to be everything to everyone," and Thiele (2016, p. 6) suggests that "Vague, hypocritical, or unsupported endorsements of sustainability may fatally weaken the concept and undermine its practice." Given the prevalence of "greenwashing" such objections are not without merit. As Grober (2012, p. 18) observes, "The most mundane of activities, even the most ruthless pillages of the planet, can be sold under the hollow label of 'sustainability' or 'sustainable development'." Sustainability has become an omni-prefix: it could be found appended to pretty much anything, at times with almost comical effects. The "most sustainable motorway ever," "sustainable golf," and a "sustainable Las Vegas" are only a few examples (ibid., p. 17). Yet the position taken here is that sustainability's polysemic nature is precisely what makes it such a rich and provocative concept. In this sense, this book seeks to elaborate on John Robinson's suggestion of a "procedural approach" to sustainability. As Robinson (2004) writes,

> sustainability can usefully be thought of as the emergent property of a con-versation about desired futures that is informed by some understanding of the ecological, social and economic consequences of different courses of action.... This view acknowledges the inherently normative and political nature of sustainability, the need for integration of different perspectives, and the recognition that sustainability is a process, not an end-state. It must be constructed through an essentially social process whereby scientific and other 'expert' information is combined with the values, preferences and beliefs of affected communities, to give rise to an emergent, 'co-produced' understanding of possibilities and preferred outcomes. (p. 381)

As will be argued and illustrated on these pages, "co-produced" under-standings of sustainability and its possibilities increasingly rely on mean-ings and imperatives that are encoded in the design, and communicated in the use of interactive media. Identifying such new meanings discloses, on the one hand, the nature of sustainability as a sociotechnical phenome-non—its unfolding as a dynamic interplay between social values and tech-nological capacities. On the other hand, it reveals the extent to which contemporary sustainability functions as a "discursive playing field" (ibid., p. 382) whose boundaries are continually pushed by new interactive technologies.

DIGITAL, NETWORKED, INTERACTIVE

As mentioned above, this book focuses on the design and use of new inter-active media. The distinction between "old" and "new" media, however, may not be as neat as some imagine it to be. Media historian Lisa Gitelman (2006), for instance, argues that not only is there nothing self-evident about the newness of new media, but that there are as many continuities as there are breaks between "old" and "new" media. Nonetheless, media scholars often point out that new media include three distinct characteristics.[8]

First, they are *digital*, that is, they are premised in the translation of continuous (analog) phenomena into discrete numeric data that can be easily adapted, changed, replicated, compressed, and stored. The shift from material indexicality to symbolic form—think of the difference between the grooves on a pressed vinyl record and the invisible bits of an MP3 file—is

---

[8] See, for instance, Flew and Smith (2011); Lister et al. (2008); Murray (2012). For an alternative view, see Manovich (2001).

what enables digital information to represent, transmediate, or "remediate" (Bolter and Grusin 1999) every other form of media. In the context of sustainability, digitization allows media creators to combine digital text with digital images, video, and sound in a manner that brings sustainability to life in vivid, dynamic, compelling, and memorable ways.[9] With the technical capacities to collate, collage, and remix at their disposal, interested individuals and environmental groups can make substantial interventions in the "representational economies" (Gershon and Bell 2013) that give meaning to environmental messaging. Digitization, however, also signals a deeper, ontological transformation. As Mark B.N. Hansen (2004) writes, digitization "underwrites a shift in the status of the medium—transforming media from forms of actual inscription of 'reality' into variable interfaces for rendering the raw data of reality" (p. 21). Ontologically speaking, new media do not just represent a preexisting reality but take part in its (re) constitution, a point to which I return below.

Second, new media are often *networked*, featuring many-to-many connections between nodes on a network, instead of the one-to-one model that underlies the telegraph and telephone, or the one-to-many model that typifies mass or broadcast media such as radio and television. The distributed network topology, enabled by packet-switching techniques and TCP/IP protocols, allows the simultaneous distribution and sharing of information by a large number of interconnected, possibly anonymous, and geographically dispersed users. It is for this reason, as briefly discussed above, that environmental communicators have found that the internet opens up exciting possibilities for disseminating environmentally significant messages while bypassing mainstream media gatekeepers, for extending the environmental public sphere virtually, and for coordinating environmental activism more effectively.

Third, and most important for the present work, new media are *interactive*, meaning that they feature feedback mechanisms that allow, invite, and even entice users to engage with them. While acknowledging that interactivity "is not a monolithic concept" (Downes and McMillan 2000, p. 159; see also Kwastek 2008), in simple terms, interactivity references the medium's capacity to facilitate reciprocal communication—allowing users to both send and receive information. However, as philosopher of

---

[9] Some of the internet's capacity for delivering rich multimedia depends on the existence of broadband connections, whose availability is an important contributor to the closing of the "digital divide."

technology Andrew Feenberg (2009) remarks, "functional objects acquire meanings that go beyond function" (p. 227). In this case, the transactional nature of interactivity is often understood as an index of participation, even empowerment. Interactivity, from this perspective, appears as "a measure of a medium's potential ability to let the user exert an influence on the content and/or form of the mediated communication" (Jensen 2008, p. 129; emphasis removed). Exceeding its strictly technical features and signaling new communicative modalities, interactivity represents "a kind of participation-enabling capability" (Andrejevic 2016, p. 199). It is in this sense that interactivity was perceived in the popular imagination as a means to erode asymmetries of communicative power, most memorably enshrined in *Time Magazine*'s celebration of Web 2.0 users as its "Person of the Year" in 2006.[10] Today such claims seem more naïve than when they were first made; however, even if interactivity on its own may not usher a new era of democratic communication, it does open up an interesting field of experience with significant effects.

While all media evoke a degree of audience activity in the form of interpretation, sense- or meaning-making, and some media even allow a certain measure of choice over content (flipping between television channels, or fast-forwarding a VCR tape, for instance),[11] by their interactivity, new media both create and cater to user expectations of being able to manipulate the actual features of the media. It is in this sense that new media give rise to what virtual reality pioneer Myron Krueger (1977/2003) calls a "responsive aesthetic." Users of new media seek and expect to be engaged in a dialectic of actions and responses that give rise to the *experience* of interactivity. Such experiences may evoke a sense of agency, or they may end up propping compulsive behavior when, for instance, interactive feedback loops are designed to increase "time-on-device" (Schüll 2013). In both cases, however, interactivity may be experienced in very different ways. As Löwgren (2009) suggests, action and response may be coupled tightly or loosely, the pace, tempo or rhythm of the back and forth may vary, the interactive flow may feature peak moments with dramatic effect, and may allow or even demand that we shift our attention between different interactive elements or from the interactive artifact to its surroundings. In every such instance,

[10] On the difference between "instrumental" and "ideological" views of interactivity, see Lister et al. (2008, pp. 21–22).

[11] Both examples manifest what Jensen (2008) terms "consultational interactivity," defined as "a measure of the media's potential ability to let the user choose, by request, from an existing selection of pre-produced information in a two-way media system with a return-channel" (p. 129).

interactivity will be experienced differently: it may result in a sense of empowerment or in frustration, in excitement or in indifference. In each and every case, however, the technical features of interactivity as a form of reciprocal communication give rise to a responsive aesthetic that unfolds as the experience of interactivity. In Feenberg's (2009) words, "Without a functional approach to experience there can be no technology, but the experience of technology is not purely functional" (p. 227).

MEDIATION

Given their particular configuration of material properties, semiotic content, and interactive form, new media evoke specific ways for users to notice, observe, and relate to other people, objects, and environments. They *mediate* the world, the entities that inhabit it, and the relations that bind them, with important consequences. Intuitively, the notion of mediation may be understood as equivalent to in-betweenness, the quality of something that stands between us and that which is external to us, a frame or window onto the world (cf. Friedberg 2006). In this, fairly naïve sense, the media (interactive or not) merely allow the world to pass through them unchanged, functioning rather passively as intermediaries that transport "meaning or force without transformation" (Latour 2005, p. 39). However, as illustrated by phenomenological approaches to media, technical mediation is a much more active, generative, and thus potentially transformative process. In Couldry and Hepp's (2017) words, our social existence, "our necessarily mediated interdependence as human beings," is "based not in some internal mental reality" but shaped by "the material processes (objects, linkages, infrastructures, platforms) through which communication, and the construction of meaning, take place" (p. 3). Winograd and Flores (1986, p. xi) give this phenomenon an existential twist: "in designing tools we are designing ways of being."

We can get a clearer sense of technological mediation from Dutch philosopher of technology Peter-Paul Verbeek's "mediation theory." Combining the insights of Don Ihde's phenomenology of technics with the conclusions of Bruno Latour's Actor-Network Theory, Verbeek (2005) argues that,

> From a hermeneutical perspective, artifacts mediate human experience by transforming perceptions and interpretive frameworks, helping to shape the way in which human beings encounter reality. The structure of this kind of mediation involves amplification and reduction; some interpretive possibilities are strength-

ened while others are weakened. From an existential perspective, artifacts mediate human existence by giving concrete shape to their behavior and the social contexts of their existence. This kind of mediation can be described in terms of translation, whose structure involves invitation and inhibition; some forms of involvement are fostered while others are discouraged. (p. 195)

In terms of perception, technologies filter, modulate, inflect, and transform our senses and experiences, amplifying certain elements of our world while attenuating others. In terms of action, technologies invite and inhibit certain behaviors. As Latour (2005) writes, "things might authorize, allow, afford, encourage, permit, suggest, influence, block, render possible, forbid, and so on" (p. 72). On both accounts, mediation leaves neither side of the "I-technology-world" relation unaffected, shaping both "how human beings can be present in the world and how the world can be present for human beings" (Verbeek 2015, p. 29).

At the same time, and as articulated persuasively in Andrew Feenberg's critical theory of technology (Feenberg 1999, 2002, 2017),[12] we should not lose sight of the fact that technological mediation is an essentially social, and therefore ideological, undertaking. As first illustrated by Marx's account of industrial machinery and later extended by the critical analyses of Herbert Marcuse, David Noble, Langdon Winner, and Feenberg, technologies embody the values, worldviews, and interests of their designers. This dynamic is captured in what Feenberg (2017) terms the "design code," the materialized translation of "worldviews and interests between the language of social actors and the technical languages of engineers or managers" (p. 74). As an analytical tool, the design code demystifies technological rationality, undermining the belief that technologies merely represent the most efficient way to do something. As Feenberg writes elsewhere,

> Technologies are selected by the dominant interests from among many possible configurations. Guiding the selection process are social codes established by the cultural and political struggles that define the horizon under which the technology will fall. Once introduced, technology offers a material validation of that cultural horizon. (Feenberg 1999, p. 87)

While the design code may not fully determine the range of possible uses and meanings a technology may carry, it reminds us that all technologies embody particular social, cultural, and political conditions.

---

[12] In his most recent book, *Technosystem: The Social Life of Reason*, Feenberg (2017) refers to his approach as "critical constructivism."

As a social enterprise, technology is never neutral. By its capacity to mediate perception and action, and as consequence of its particular design code, every technology affects the meaning of the activities in which it is implicated. Terry Winograd and Fernando Flores (1986) explain that when users interact with technology, they are guided by certain background suppositions and expectations that inform their understanding of what the technology does, how, and with what consequences. These background assumptions, part of what Gadamer (1975/2004) calls the "hermeneutic horizon" and Jauss (1982) calls the "horizon of expectations," are produced by specific traditions, histories, and social conditions, and are influenced by users' previous experiences with technology.

The particular way in which the interaction unfolds may confirm these assumptions, or call them into question. In the event of the latter, the interaction may potentiate new meanings and new possibilities for action that inform future technological interactions. In other words, when we interact with technologies, we construct *in dialogue* new ways to not only act with technology but also to make sense of the entire nexus of practices, significances, and meanings that technology relates. Technological mediation, it follows, is a form of *meaning-giving*. Similar to what Wittgenstein (1953/2001) describes as "language games," and Coeckelbergh (2017) calls "technology games," the meaning of technology, and of technologically mediated activities, changes in practice. Old meanings may disappear, and new meanings may emerge. As we engage with the world through interactive technologies, the very nature of the medium affects and pluralizes the kind of meanings we derive from the activity.

As noted above, deriving and attributing meaning does not take place in a vacuum, but is influenced by previous experiences, cultural conditions, social conventions, and political forces. It is therefore a circular process, with no clear beginning nor end, that may yield unanticipated consequences. As McLuhan (1962) shows, mediation percolates and sediments into culture, but culture influences the range of effects mediations may carry.[13] Couldry (2008) describes this circularity as follows:

> 'media' work, and must work, not merely by transmitting discrete textual units for discrete moments of reception, but through a process of environmental transformation which, in turn, transforms the conditions under

---

[13] "Every culture and every age has its favorite model of perception and knowledge that it is inclined to prescribe for everybody and everything" (McLuhan 1964, p. 5). The cumulative outcomes of acts of mediation have been described recently as a process of "mediatization" (see, for instance, Couldry 2008; Couldry and Hepp 2017; Hjarvard 2013).

which any future media can be produced and understood. In other words, 'mediation' is a non-linear process. (p. 380)

We change with every technical interaction, and so does the world, even if those changes are often quite nuanced and hard to detect without significant reflection. Nonetheless, if we assume, as Alexander Galloway (2012) does, that "flows of signification organize a certain knowledge of the world and a commitment to it" (p. 45), tracing those flows and their consequences is key to understanding how the design and use of interactive media influences the perception and meaning of contemporary sustainability.

A Brief Outline of the Book's Structure

The breadth and depth of the book's topic calls for a trans-, inter-, or to borrow Robinson's (2008) provocative suggestion, an un-disciplinary approach.[14] The analysis of the interrelated functionality and meaning of interactive media presented here draws from a wide array of sources that include contemporary work in human-computer interaction (HCI), social theory, media studies, and the philosophy of technology. It is interpretive in nature, loosely following what Blevis et al. (2007) describe as "design critique": "*a process of discourse on many levels of the nature and effects of an ultimate particular design....* The role of the design critic is to comment on the qualities of an ultimate particular from an holistic perspective, including reason, ethics, and aesthetics as well as minute details of form and external effects on culture" (p. 24; emphasis in origin).[15] There is some similarity here with "close reading" techniques used in the study of literature: "a technically informed, fine-grained analysis of some piece of writing, usually in connection with some broader question of interest" (Herrnstein Smith 2015, p. 3). But whereas close readings commence with the written word, design critiques commence with the affordances of interactive media, that is, "the perceived and actual properties of the thing, primarily those fundamental properties that determine just how the thing could possibly be used" (Norman 2002, p. 19).

---

[14] Robinson, to be clear, does not suggest research be undisciplined but that interdisciplinarity be issue-driven. For a more recent take on the topic, see Darbellay (2015).

[15] An "ultimate particular" could be any specific technological artifact or service. The term is meant to draw attention to the notion that a technology could be at once mass produced (and standardized), and uniquely situated in terms of its use context (see also Stolterman 2008).

Affordances can be thought of as signifiers of possible actions (Norman 2011); however, they are not independent, intrinsic features of technical objects but are always affordances *for someone*. They are a "relationship" (ibid., p. 228). As Feenberg (2017) explains, "The chair has a function as a thing on which to sit only in so far as it is recognized as a chair, that is to say, only in so far as its meaning is apprehended by potential users. Assignments of meaning depend on a human subject" (p. 28). As latent potentialities that become articulated in practice, affordances appear as material-perceptual anchors for sociotechnical meaning-making and meaning-giving. They emerge as (and encode) a relation between users, technical artifacts, and the world. Given the dynamic nature of this relation, no single interpretation can capture what a technical artifact means to all of its actual (and potential) users. Yet some interpretations can be more plausible or valuable than others. For that reason, and where possible, the interpretive analysis is complemented by accounts of user experiences with the media, and interviews with media designers.[16] These accounts, along with the more "technically informed" analysis of the media themselves, animate the book's central argument: *the meaning of sustainability is refracted as consequence of its increased mediation by interactive technologies.*

The book presents four such refracted meanings in four chapters, each pertaining to a distinct domain of design activity:

Chapter 2 describes the design and use of interactive media aimed at closing the "attitude-behavior gap" (Kollmuss and Agyeman 2002) by influencing the behaviors of their users. The chapter draws from contemporary work in cognitive science and behavioral psychology and surveys a wide variety of persuasive interactions with varying degrees of prescriptiveness—from "seducing" users (Tromp et al. 2011) to engaging them with "technologies of behavior" (Skinner 1971). The image of sustainability that emerges from persuasive uses of interactive media is of a delicate balance between human and nonhuman interests, one that was disturbed by modern production and consumption but can be restored by inculcating more sustainable, or less unsustainable, behaviors.

---

[16] Importantly, the meanings users derive from their technologically mediated interactions are never limited to what designers intend. Different interpretations and possibilities for appropriation are inherent to every technology. For an influential account of technology's "interpretative flexibility," see Pinch and Bijker (1984). On appropriative uses of technology, see the essays collected in Eglash et al. (2004). And for implications for design, see Redström (2008).

Chapter 3 focuses on the design and use of interactive media to scaffold their users' understanding of sustainability. The chapter traces the influence of concepts developed in complex systems theory on the design of sustainability games and simulations, and asks about the implications of using computational assemblages to represent and act on socio-natural systems. The image of sustainability that emerges from these media and the synoptic interactions they provide is of a complex problem that could be solved, or at least managed, by applying scientific methods and reasoning.

Chapter 4 describes the design and use of immersive media to evoke resonant, emotionally stirring experiences of sustainability. The analysis draws from phenomenological conceptions of presence to illustrate how virtual reality (VR) games and interactive installations create situations that allow users to experience sustainability (and unsustainability) in embodied, affective ways. Through the perspective provided by the resonant interactions such media evoke, sustainability is seen as felt embeddedness, a field of interrelated phenomena that need to be brought closer to users' everyday life.

Chapter 5 focuses on interactive media that promote imaginative engagement with sustainability. By deploying what I call "worldmaking interactions" (Bendor 2018), the interactive media discussed in this chapter seek to cultivate their users' capacity to imagine and pursue alternative social, scientific, and political configurations. Sustainability emerges from such media as an imaginary, a means by which deeply held beliefs about the world can be made visible and legible, can be evaluated, and perhaps even transformed.

The order of the book's chapters is not incidental. It reflects my belief that for sustainability to deliver on its transformative promises, it must critically engage with social, political, and economic structures. So while there is certainly value in promoting and supporting the kind of sustainable behaviors and lifestyles discussed in Chap. 2, I have little doubt that the key to the kind of deep changes we so desperately need lies not in the cumulative consumptive behaviors of individuals but in the capacity of individuals to collectively organize and make strategic interventions in politics—to become sustainability citizens (Nelson 2016). How such interventions may transpire, and how their structural transformations may be achieved, are topics so broad, complex, and contentious that they merit a different kind of evaluation, one that is beyond the scope of the present work.[17] The order of the chapters, however, serves as an affirmation that

---

[17] Those wishing to engage with the topic more deeply may find useful points of departure in Klein (2014), and, more recently, Bonneuil and Fressoz (2017) and Mann and Wainwright (2018).

deep, structural changes start with reasonable exchanges, gain momentum with experiential resonance, and become truly transformative when they seek and succeed to affect social and political imaginaries.

The reader may find that the four meanings of sustainability discussed on these pages differ significantly, and may exhibit varying degrees of incompatibility in terms of their assumptions about the nature of humans, the essence of the world, how society can become more sustainable, and how interactive technologies could facilitate the process. Yet, in practice, the different media *forms* implicated with these meanings may be combined to produce multifaceted, more holistic interactive experiences. Persuasive interactions may appear alongside resonant ones, just as synoptic interactions may serve to prepare or prime users to engage with world-making interactions. As consequence, sustainability may gain additional, unanticipated meanings. Rather than invalidating the account given here, I suggest that this capacity to be refracted is further indication of sustainability's richness and value, a promise that future media will unfold even more ways for sustainability to become meaningful and transformative.

Whether it subscribes to a single meaning of sustainability, combines a few, or presents new hermeneutic permutations, every medium has consequences, and these, even as mere potentials, merit critical evaluation. The book's concluding chapter attempts such an evaluation based on three elements: complexity, futurity, and agency. Complexity because sustainability involves a dense network of human and nonhuman actors, material and cultural imperatives whose interactions often lead to unanticipated, emergent consequences. Futurity because from its very beginning, sustainability sought to shift human temporal horizons, considerations, and responsibilities from the "here and now" to the future. And agency because sustainability implies that humans have the capacity to intervene and make the world a better place. In writing this book, my goal is to help increase our capacity to act with, through, and on sustainability. Our prospects for surviving the Anthropocene lie in the balance.

Bibliography

Anderson, A. G. (2014). *Media, Environment and the Network Society*. Basingstoke/ New York: Palgrave.

Andrejevic, M. (2016). The Pacification of Interactivity. In D. Barney, G. Coleman, C. Ross, J. Sterne, & T. Tembeck (Eds.), *The Participatory Condition in the Digital Age* (pp. 187–206). Minneapolis/London: University of Minnesota Press.

Atanasova, D., & Koteyko, N. (2017). Metaphors in Guardian Online and Mail Online Opinion-Page Content on Climate Change: War, Religion, and Politics. *Environmental Communication, 11*(4), 452–469.

Bendor, R. (2018). Interaction Design for Sustainability Futures: Towards Worldmaking Interactions. In M. Hazas & L. P. Nathan (Eds.), *Digital Technology and Sustainability: Engaging the Paradox* (pp. 205–216). New York: Routledge.

Blevis, E. (2007). Sustainable Interaction Design: Invention & Disposal, Renewal & Reuse. In M. B. Rosson & D. Gilmore (Eds.), *Proceedings of CHI 2007* (pp. 503–512). New York: ACM Press.

Blevis, E., Lim, Y.-k., Roedl, D., & Stolterman, E. (2007). Using Design Critique as Research to Link Sustainability and Interactive Technologies. In D. Schuler (Ed.), *Online Communities and Social Computing* (pp. 22–31). Berlin/Heidelberg: Springer.

Bolter, J. D., & Grusin, R. A. (1999). *Remediation: Understanding New Media.* Cambridge, MA: MIT Press.

Bonneuil, C., & Fressoz, J.-B. (2017). *The Shock of the Anthropocene.* London/New York: Verso.

Büscher, B. (2014). Nature 2.0: Exploring and Theorizing the Links Between New Media and Nature Conservation. *New Media & Society, 18*(5), 726–743.

Busse, D., Mann, S., Nathan, L. P., & Preist, C. (2013). Changing Perspectives on Sustainability: Healthy Debate or Divisive Factions. In *CHI 2013 Extended Abstracts on Human Factors in Computing Systems* (pp. 2505–2508). New York: ACM.

Caradonna, J. L. (2014). *Sustainability: A History.* New York: Oxford University Press.

Clear, A., Preist, C., Joshi, S., Nathan, L. P., Mann, S., & Nardi, B. A. (2015). Expanding the Boundaries: A SIGCHI HCI & Sustainability Workshop, CHI '15 Extended Abstracts on Human Factors in Computing Systems (pp. 2373–2376). New York: ACM.

Coeckelbergh, M. (2017). *Using Words and Things: Language and Philosophy of Technology.* London/New York: Routledge.

Connelly, S. (2007). Mapping Sustainable Development as a Contested Concept. *Local Environment: The International Journal of Justice and Sustainability, 12*(3), 259–278.

Couldry, N. (2008). Mediatization or Mediation? Alternative Understandings of the Emergent Space of Digital Storytelling. *New Media & Society, 10*(3), 373–391.

Couldry, N., & Hepp, A. (2017). *The Mediated Construction of Reality.* Cambridge/Malden: Polity Press.

Cox, J. R., & Pezzullo, P. C. (2015). *Environmental Communication and the Public Sphere* (4th ed.). Thousand Oaks: Sage.

Darbellay, F. (2015). Rethinking Inter- and Transdisciplinarity: Undisciplined Knowledge and the Emergence of a New Thought Style. *Futures, 65,* 163–174.

Dijksterhuis, E. J. (1961). *The Mechanization of the World Picture* (trans Dikshoorn, C.). London/New York: Oxford University Press.

DiSalvo, C., Sengers, P., & Brynjarsdóttir, H. (2010). Mapping the Landscape of Sustainable HCI. In *Proceedings of CHI 2010* (pp. 1975–1984). New York: ACM Press.

Dourish, P. (2010). HCI and Environmental Sustainability: The Politics of Design and the Design of Politics. In O. W. Bartelsen & P. Krogh (Eds.), *Proceedings of DIS 2010* (pp. 1–10). New York: ACM.

Downes, E. J., & McMillan, S. (2000). Defining Interactivity: A Qualitative Identification of Key Dimensions. *New Media & Society, 2*(2), 157–179.

Eglash, R., Croissant, J., Di Chiro, G., & Fouché, R. (Eds.). (2004). *Appropriating Technology: Vernacular Science and Social Power.* Minneapolis: University of Minnesota Press.

Ehrenfeld, J. (2008). *Sustainability by Design: A Subversive Strategy for Transforming Our Consumer Culture.* New Haven: Yale University Press.

Farley, H. M., & Smith, Z. A. (2014). *Sustainability: If It's Everything, Is It Nothing?* London/New York: Routledge.

Feenberg, A. (1999). *Questioning Technology.* London/New York: Routledge.

Feenberg, A. (2002). *Transforming Technology: A Critical Theory Revisited.* New York: Oxford University Press.

Feenberg, A. (2009). Peter-Paul Verbeek: Review of *What Things Do. Human Studies, 32*(2), 225–228.

Feenberg, A. (2017). *Technosystem: The Social Life of Reason.* Cambridge, MA: Harvard University Press.

Figueres, C., Schellnhuber, H. J., Whiteman, G., Rockström, J., Hobley, A., & Rahmstorf, S. (2017). Three Years to Safeguard Our Climate. *Nature, 546,* 593–595.

Flew, T., & Smith, R. (2011). *New Media: An Introduction (Canadian Edition)* (2nd ed.). Oxford/New York: Oxford University Press.

Frick, T. (2016). *Designing for Sustainability: A Guide to Building Greener Digital Products and Services.* Sebastopol: O'Reilly.

Friedberg, A. (2006). *The Virtual Window: from Alberti to Microsoft.* Cambridge, MA: MIT Press.

Fritsch, J., & Brynskov, M. (2011). Between Experience, Affect, and Information: Experimental Urban Interfaces in the Climate Change Debate. In M. Foth, L. Forlano, C. Satchell, & M. Gibbs (Eds.), *From Social Butterfly to Engaged Citizen: Urban Informatics, Social Media, Ubiquitous Computing, and Mobile*

Technology to Support Citizen Engagement (pp. 115–134). Cambridge, MA: MIT Press.

Fuad-Luke, A. (2009). *Design Activism: Beautiful Strangeness for a Sustainable World*. London/Sterling: Earthscan.

Funk, J. (2017). Assessing Public Forecasts to Encourage Accountability: The Case of MIT's Technology Review. *PLoS One, 12*(8), e0183038.

Gadamer, H. G. (2004). *Truth and Method* (trans: Weinsheimer, J., & Marshall, D. G., 3rd ed.). New York/London: Continuum.

Galloway, A. R. (2012). *The Interface Effect*. Malden/Cambridge: Polity.

Gershon, I., & Bell, J. A. (2013). Introduction: The Newness of New Media. *Culture, Theory and Critique, 54*(3), 259–264.

Gitelman, L. (2006). *Always Already New: Media, History and the Data of Culture*. Cambridge, MA: MIT Press.

Grober, U. (2012). *Sustainability: A Cultural History* (trans: Cunningham, R.). Totnes: Green Books.

Haider, J. (2016). The Shaping of Environmental Information in Social Media: Affordances and Technologies of Self-Control. *Environmental Communication, 10*(4), 473–491.

Hansen, M. B. N. (2004). *New Philosophy for New Media*. Cambridge, MA: MIT Press.

Hansen, A. (2010). *Environment, Media and Communication*. London/New York: Routledge.

Hansen, A., & Cox, J. R. (Eds.). (2015). *The Routledge Handbook of Environment and Communication*. London/New York: Routledge.

Hazas, M., & Nathan, L. P. (2018a). Introduction: Digital Technology and Sustainability: Engaging the Paradox. In M. Hazas & L. P. Nathan (Eds.), *Digital Technology and Sustainability: Engaging the Paradox* (pp. 3–13). New York: Routledge.

Hazas, M., & Nathan, L. P. (Eds.). (2018b). *Digital Technology and Sustainability: Engaging the Paradox*. New York: Routledge.

Heidegger, M. (1977). *The Question Concerning Technology, and Other Essays* (trans: Lovitt, W., 1st ed.). New York: Harper & Row.

Herrnstein Smith, B. (2015). *What Was "Close Reading"? A Century of Method in Literary Studies*. Paper Presented at the Heyman Center, Columbia University, New York, for a Digital Humanities Workshop series "On Method".

Hilty, L. M., & Aebischer, B. (Eds.). (2015). *ICT Innovations for Sustainability*. Cham: Springer.

Hjarvard, S. (2013). *The Mediatization of Culture and Society*. Abington/New York: Routledge.

IPCC. (2015). *Climate Change* 2014: *Synthesis Report. Contribution of Working Groups I, II and III to the Fifth Assessment Report of the Intergovernmental*

Panel on Climate Change. Retrieved from Geneva, https://www.ipcc.ch/pdf/
assessment-report/ar5/syr/SYR_AR5_FINAL_full_wcover.pdf

Issa, T., Isaias, P., & Issa, T. (Eds.). (2017). *Sustainability, Green IT and Education Strategies in the Twenty-First Century.* Cham: Springer.

Jacobs, M. (1999). Sustainable Development as a Contested Concept. In A. Dobson (Ed.), *Fairness and Futurity: Essays on Environmental Sustainability and Social Justice* (pp. 21–45). Oxford/New York: Oxford University Press.

Jauss, H. R. (1982). *Toward an Aesthetic of Reception* (trans: Bahti, T.). Minneapolis: University of Minnesota Press.

Jensen, J. F. (2008). *The Concept of Interactivity – Revisited: Four New Typologies for a New Media Landscape.* Paper Presented at uxTV 2008, Silicon Valley.

Katz-Kimchi, M., & Manosevitch, I. (2015). Mobilizing Facebook Users Against Facebook's Energy Policy: The Case of Greenpeace Unfriend Coal Campaign. *Environmental Communication, 9*(2), 248–267.

Kelly, M. R. (2012). Climate Change Communication Research Here and Now: A Reflection on Where We Came From and Where We Are Going. *Applied Environmental Education & Communication, 11*(3–4), 117–118.

Kirilenko, A. P., & Stepchenkova, S. O. (2014). Public Microblogging on Climate Change: One Year of Twitter Worldwide. *Global Environmental Change, 26,* 171–182.

Klein, N. (2014). *This Changes Everything: Capitalism vs. the Climate.* Toronto: Alfred A. Knopf.

Knowles, B., Blair, L., Coulton, P., & Lochrie, M. (2014). Rethinking Plan A for Sustainable HCI. In *Proceedings of CHI 2014* (pp. 3593–3596). New York: ACM.

Kollmuss, A., & Agyeman, J. (2002). Mind the Gap: Why Do People Act Environmentally and What Are the Barriers to Pro-environmental Behavior? *Environmental Education Research, 8*(3), 239–260.

Krueger, M. W. (1977/2003). Responsive Environments. In N. Wardrip-Fruin & N. Montfort (Eds.), *The New Media Reader* (pp. 379–389). Cambridge, MA/London: MIT Press.

Kumar, K. (1978). *Prophecy and Progress: The Sociology of Industrial and Post-Industrial Society.* Harmondsworth/New York: Penguin.

Kwastek, K. (2008). Interactivity – A Word in Process. In C. Sommerer, L. C. Jain, & L. Mignonneau (Eds.), *The Art and Science of Interface and Interaction Design* (pp. 15–26). Berlin & Heidelberg: Springer.

Latour, B. (2005). *Reassembling the Social: An Introduction to Actor-Network-Theory.* Oxford/New York: Oxford University Press.

Leiss, W. (1972). *The Domination of Nature.* New York: G. Braziller.

Lewis, S. L., & Maslin, M. A. (2015). Defining the Anthropocene. *Nature, 519*(7542), 171–180.

Lister, M., Dovey, J., Giddings, S., Grant, I., & Kelly, K. (2008). *New media: A Critical Introduction* (2nd ed.). Abingdon/New York: Routledge.

Löwgren, J. (2009). Toward an Articulation of Interaction Esthetics. *New Review of Hypermedia and Multimedia, 15*(2), 129–146.

Mann, G., & Wainwright, J. (2018). *Climate Leviathan: A Political Theory of Our Planetary Future.* London/New York: Verso.

Manovich, L. (2001). *The Language of New Media.* Cambridge, MA: MIT Press.

McGinn, R. E. (1990). What Is Technology? In L. A. Hickman (Ed.), *Technology as a Human Affair* (pp. 10–25). New York: McGraw Hill.

McLuhan, M. (1962). *The Gutenberg Galaxy; The Making of Typographic Man.* Toronto: University of Toronto Press.

McLuhan, M. (1964). *Understanding Media; The Extensions of Man.* New York: McGraw-Hill.

Merchant, C. (1989). *The Death of Nature: Women, Ecology, and the Scientific Revolution.* New York: Harper & Row.

Moser, S. C. (2010). Communicating Climate Change: History, Challenges, Process and Future Directions. *Wiley Interdisciplinary Reviews: Climate Change, 1*(1), 31–53.

Moser, S. C. (2016). What More Is There to Say? Reflections on Climate Change Communication Research and Practice in the Second Decade of the 21st Century. *Wiley Interdisciplinary Reviews: Climate Change, 7*(3), 345–369.

Moser, S. C., & Dilling, L. (Eds.). (2007). *Creating a Climate for Change: Communicating Climate Change and Facilitating Social Change.* Cambridge/New York: Cambridge University Press.

Murray, J. H. (2012). *Inventing the Medium: Principles of Interaction Design as a Cultural Practice.* Cambridge, MA: MIT Press.

Nardi, B., & Ekebia, H. (2018). Developing a Political Economy Perspective for Sustainable HCI. In M. Hazas & L. P. Nathan (Eds.), *Digital Technology and Sustainability: Engaging the Paradox* (pp. 86–102). New York: Routledge.

Nelson, A. (2016). The Praxis of Sustainability Citizenship. In R. Horne, J. Fien, B. B. Beza, & A. Nelson (Eds.), *Sustainability Citizenship in Cities: Theory and Practice* (pp. 17–28). London/New York: Routledge.

Norman, D. A. (2002). *The Design of Everyday Things* (2nd ed.). New York: Doubleday.

Norman, D. A. (2011). *Living with Complexity.* Cambridge, MA: MIT Press.

O'Neill, S., & Boykoff, M. (2011). The Role of New Media in Engaging the Public with Climate Change. In L. Whitmarsh, S. O'Neill, & I. Lorenzoni (Eds.), *Engaging the Public with Climate Change: Behaviour Change and Communication* (pp. 233–251). London/Washington, DC: Earthscan.

Ortega y Gasset, J. (1941). *Toward a Philosophy of History.* New York: W.W. Norton.

Pearce, W., Holmberg, K., Hellsten, I., & Nerlich, B. (2014). Climate Change on Twitter: Topics, Communities and Conversations About the 2013 IPCC Working Group 1 Report. *PLoS One, 9*(4), e94785.

Pinch, T. J., & Bijker, W. E. (1984). The Social Construction of Facts and Artefacts: Or How the Sociology of Science and the Sociology of Technology Might Benefit Each Other. *Social Studies of Science, 14*, 399–441.

Pogue, D. (2012, January 18). Use It Better: The Worst Tech Predictions of All Time. *Scientific American*. Retrieved from https://www.scientificamerican.com/article/pogue-all-time-worst-tech-predictions

Prost, S., Schrammel, J., & Tscheligi, M. (2014). 'Sometimes It's the Weather's Fault': Sustainable HCI & Political Activism, CHI '14 Extended Abstracts on Human Factors in Computing Systems (pp. 2005–2010). New York: ACM.

Redström, J. (2008). RE: Definitions of Use. *Design Issues, 29*(4), 410–423.

Robinson, J. (2004). Squaring the Circle? Some Thoughts on the Idea of Sustainable Development. *Ecological Economics, 48*, 369–384.

Robinson, J. (2008). Being Undisciplined: Transgressions and Intersections in Academia and Beyond. *Futures, 40*(1), 70–86.

Rockström, J. (2015, November 14). The Planet's Future Is in the Balance. But a Transformation Is Already Under Way. *The Guardian*. Retrieved from https://www.theguardian.com/environment/2015/nov/14/un-climate-change-summit-paris-planet-future-balance-science

Schäfer, M. S. (2012). Online Communication on Climate Change and Climate Politics: A Literature Review. *Wiley Interdisciplinary Reviews: Climate Change, 3*(6), 527–543.

Schüll, N. D. (2013). *Addiction by Design: Machine Gambling in Las Vegas.* Princeton/Oxford: Princeton University Press.

Schwab, K. (2017). *The Fourth Industrial Revolution.* New York: Crown Business.

Silberman, M. S., Blevis, E., Huang, E., Nardi, B. A., Nathan, L. P., Busse, D., et al. (2014a). What Have We Learned? A SIGCHI HCI & Sustainability Community Workshop. *CHI '14 Extended Abstracts on Human Factors in Computing Systems* (pp. 143–146). New York: ACM.

Silberman, M. S., Knowles, B., Nathan, L., Bendor, R., Clear, A., Hakansson, M., et al. (2014b). Next Steps for Sustainable HCI. *Interactions, 21*(5), 66–69.

Skarda, E. (2011, October 21). Top 10 Failed Predictions: Technology? What's That? *Time Magazine*. Retrieved from http://content.time.com/time/specials/packages/article/0,28804,2097462_2097456_2097467,00.html

Skinner, B. F. (1971). *Beyond Freedom and Dignity.* Indianapolis/Cambridge: Hackett.

Spartz, J. T., Su, L. Y.-F., Griffin, R., Brossard, D., & Dunwoody, S. (2017). YouTube, Social Norms and Perceived Salience of Climate Change in the American Mind. *Environmental Communication, 11*(1), 1–16.

Starosielski, N., & Walker, J. (Eds.). (2016). *Sustainable Media: Critical Approaches to Media and Environment*. New York: Routledge.

Stephens, S. H., DeLorme, D. E., & Hagen, S. C. (2017). Evaluation of the Design Features of Interactive Sea-Level Rise Viewers for Risk Communication. *Environmental Communication, 11*(2), 248–262.

Stolterman, E. (2008). The Nature of Design Practice and Implications for Interaction Design Research. *International Journal of Design, 2*(1), 55–65.

Thiele, L. P. (2016). *Sustainability* (2nd ed.). Cambridge/Malden: Polity Press.

Tomlinson, B. (2010). *Greening Through IT: Information Technology for Environmental Sustainability*. Cambridge, MA: MIT Press.

Tromp, N., Hekkert, P., & Verbeek, P.-P. (2011). Design for Socially Responsible Behavior: A Classification of Influence Based on Intended User Experience. *Design Issues, 27*(3), 3–19.

Verbeek, P.-P. (2005). *What Things Do: Philosophical Reflections on Technology, Agency, and Design* (trans: Crease, R. P.). University Park: Pennsylvania State University Press.

Verbeek, P.-P. (2015). Beyond Interaction: A Short Introduction to Mediation Theory. *Interactions, 12*(3), 26–31.

Winograd, T., & Flores, F. (1986). *Understanding Computers and Cognition: A New Foundation for Design*. Norwood: Ablex Pub. Corp.

Wittgenstein, L. (2001). *Philosophical Investigations* (trans: Anscombe, G. E. M., 3rd ed.). Oxford/Malden: Blackwell.

CHAPTER 2

Behavior

SUSTAINABLE BEHAVIOR BY DESIGN

In anticipation of the 2015 edition of Clean Cornwall week in Cornwall, United Kingdom, local organizers conducted a small experiment. They placed a green colored plastic bag conspicuously on a pedestrian pathway, and in close proximity to a garbage bin. In two hours, and in two different locations, 266 people passed right next to the bag. Only two stopped to pick it up and throw it in the bin. Indeed, in the video that documents the experiment, some passersby are seen deliberately walking around or even over the litter in an effort to avoid the unpleasant task.[1] This unflattering rate of action (under 1%) becomes even more ironic when considering that in a poll conducted near the time of the experiment, 94% of respondents said that litter was a problem in Cornwall. So would Cornwallians please make up their mind: do they want a cleaner environment or not? And if they do, are they willing to actually do something about it?

As tempting as it may be to single out the citizens of Cornwall for their lack of pro-environmental behavior, similar findings were documented across a range of experiments in social psychology as early as 1971 (McKenzie-Mohr 2000). Reporting on what is perhaps the most famous experiment in the social psychology of environmental behavior, Bickman

---

[1] See report and video here: https://www.businesscornwall.co.uk/news-by-location/truro-business-news/2015/09/video-clean-cornwall-week (last accessed Mar. 18, 2018).

© The Author(s) 2018
R. Bendor, *Interactive Media for Sustainability*,
Palgrave Studies in Media and Environmental Communication,
https://doi.org/10.1007/978-3-319-70383-1_2

(1972) writes that only 8 of 506 (1.4%) people on Smith College campus who passed by litter directly placed in their path stopped to pick it up. A fifth of those who neglected to pick up the litter were interviewed shortly after, and of those, in eerie similarity to Cornwall, 94% agreed that it was "everyone's responsibility to pick up litter when they see it" (p. 324).

Although they are separated by almost 45 years, during which the environmental movement has grown into a global environmental consciousness, both the Cornwall and the Smith College experiments indicate a palpable gap between people's awareness of and willingness to act on environmental issues and their actual behavior (or lack thereof). Kollmuss and Agyeman (2002) call this the "attitude-behavior gap." We like to talk the talk, as they say, but seem much less capable of walking the walk. But what if it was possible to override cognitive dissonance and encourage people to behave pro-environmentally *by design*? What if, for instance, designers could transform the unappealing, mundane act of garbage disposal into a more fun or satisfying activity? Would the public be more likely to act on their intentions and "do the right thing"? The folks at Volkswagen thought so. As part of its Fun Theory project, the company created "the world's deepest bin": a garbage disposal bin that generates a long shrieking sound for each article deposited in it, giving the illusion that the bin was much deeper than it actually was.[2] The result, Volkswagen states, was that the bin collected 72 kg of rubbish in one day, a whopping 41 kg more than another, "regular" bin located nearby. "Fun can obviously change behaviour for the better," the company writes. But so can social incentives and pressures, or at least that was the thought behind the BinCam, a small, mobile phone-based camera and app that can be installed in domestic garbage bins. The camera snaps images of all the objects discarded in the bin, which the app then analyzes and posts on social media as a means to raise awareness of household disposal practices and stir recycling competitions between households. In a paper reporting on the device's experimental run, its designers state that it inspired "an increase in individuals' awareness, reflection and perceived behavioral control related to their waste behavior" (Thieme et al. 2012, p. 2342).

Both Volkswagen's "deepest bin" and the BinCam make evident the kind of creative solutions designers attempt as a means to close the attitude-behavior gap. But they also hint at the kind of criticisms such interventions attract. The "deepest bin" seems more a public relations

---

[2] See http://www.thefuntheory.com (last accessed Mar. 18, 2018).

campaign than a scalable solution, and it is unclear whether the bin has indeed increased the amount of garbage properly disposed or merely attracted disposal from other bins. BinCam raises concerns over privacy, the ethics, and utility of social behavioral cues, and about the overall effectiveness of micro-behavioral interventions. Nonetheless, the two provide a good starting point to consider persuasive design: an umbrella term that refers to a range of design strategies that aim to change behaviors without coercion.[3] When applied to interactive media, persuasive design manifests as *persuasive interactions*. The subject of this chapter is the application of persuasive interactions to promote behavioral change in support of sustainability.

MIND THE GAP(S)

The attitude-behavior gap described by Kollmuss and Agyeman (2002) may take several forms, each reflecting a different kind of challenge. Take, for example, Michael Pollan's observation in the *New York Times*'s Green Issue from April 20, 2008:

> Let's say I do bother, big time. I turn my life upside-down, start biking to work, plant a big garden, turn down the thermostat so low I need the Jimmy Carter signature cardigan, forsake the clothes dryer for a laundry line across the yard, trade in the station wagon for a hybrid, get off the beef, go completely local. I could theoretically do all that, but what would be the point when I know full well that halfway around the world there lives my evil twin, some carbon-footprint *doppelgänger* in Shanghai or Chongqing who has just bought his first car (Chinese car ownership is where ours was back in 1918), is eager to swallow every bite of meat I forswear and who's positively itching to replace every last pound of CO_2 I'm struggling no longer to emit. So what exactly would I have to show for all my trouble?

The context of Pollan's gripe has changed considerably following the United States and China's agreement to cooperate on climate change

---

[3] What will be referred to here as "persuasive design" includes (among others) "persuasive systems design," "design with intent," "affective computing," "persuasive technologies," "design for sustainable behavior change," "persuasive sustainability systems," and "behavior change support systems." For overviews of the different approaches and how they compare, see Brynjarsdóttir et al. (2012), Niedderer et al. (2016), Oinas-Kukkonen (2013), Zachrisson and Boks (2012), Zapico et al. (2009).

mitigation and clean energy development,[4] and with the latest measures introduced by the Chinese government to phase out fossil-fueled vehicles.[5] It is more likely that it is Pollan's "*doppelgänger* in Shanghai or Chongqing" that now feels exasperated by the United States's lack of action on climate change. But the dilemma Pollan expresses has not lost its relevance: how can one be moved to act when the issue—often a "bigger-than-self" problem (Crompton 2010)—can only be addressed cumulatively by the combined efforts of millions? Furthermore, when human impact on the environment is measured in geological terms, as implied by the recent naming of our current epoch the Anthropocene, what can one really do to make a difference? As well-meaning as we may be, small micro-activities seem woefully inadequate to tackle our global, civilizational crisis.

The magnitudinal gap—the seemingly unbridgeable chasm between the immense scale of our environmental problems and our meager capacity to effectively act on them—may be just too wide to bridge without a considerable leap of faith. But it is hardly the only gap that seems immune to our well intentions: there are gaps between the possession of necessary and available information, between perceived individual and social benefits of a particular behavior, between perceived ease and real difficulty to act, between individual values and societal norms, between expectations for immediate results and time-delayed outcomes—the litany of gaps only seems to grow, while our capacity to deal with them seems to diminish. When translated into the language of psychologist Albert Bandura (1994), many such variations on the attitude-behavior gap indicate a troubling deficit of "perceived self-efficacy": on the one hand, people may lack the conviction that they have what it takes to deal with situations of this magnitude. On the other hand, people may simply be unaware of how they may actually achieve their goals. Looking at pro-environmental behavior through the prism of self-efficacy, Bamberg and Möser (2007) conclude that,

> on average, the intention to perform a pro-environmental behavioural option can be described as a weighted balance of information concerning the three questions "How many positive/negative personal consequences would result from choosing this pro-environmental option compared to

[4] https://obamawhitehouse.archives.gov/the-press-office/2014/11/11/fact-sheet-us-china-joint-announcement-climate-change-and-clean-energy-c (last accessed Mar. 18, 2018).
[5] https://www.bloomberg.com/news/articles/2017-09-10/china-s-fossil-fuel-deadline-shifts-focus-to-electric-car-race-j7fktx9z (last accessed Mar. 18, 2018).

other options?", "How difficult would be the performance of the pro-environmental option compared to other options?", and "Are there reasons indicating a moral obligation for performing the pro-environmental option?" (p. 21)

Never has the decision to pick up a green bag and put it in the nearby bin appeared so complicated. But must it be?

Early attempts to understand and influence pro-environmental behavior relied on a fairly linear model of human behavior, according to which possessing knowledge of an environmental issue was considered the main predictor of action on it (see Image 2.1). Given that to become knowledgeable about an issue one first has to become aware of it and then develop a concern about it, providing compelling information about an issue was expected to create the attitudes that would, in turn, spur appropriate behavior. Thus, in what came to be known as the information deficit or the "diffusionist" model, the expert provision of appropriate, relevant information, most often drawn from scientific sources, became the central aim of science communication and, by extension, environmental communication (Burgess et al. 1998; Gregory and Miller 1998). As Bucchi (2008) points out, belief in the overwhelming complexity of scientific knowledge and the inability of "regular folks" to comprehend modern science—on display, for instance, in Einstein's famous statement that only a dozen people may understand his General Theory of Relativity—prompted a view of the media as indispensable translators of scientific insight. If the public suffered a knowledge deficit, the media will fill it. The result was a decontextualized, top-down, one-size-fits-all, "linear, pedagogical and paternalistic view of communication" (ibid., p. 58).

In the context of promoting sustainable behavior, the information deficit model often operates in tandem with the economic rationality model (or its somewhat improved successor, the bounded rationality model). The theory of economic rationality holds that given sufficient information about a situation, humans will always act in their best economic interests,

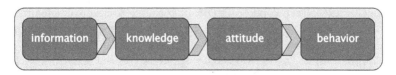

Image 2.1   Schematic view of the information deficit model

that is, in ways that maximize benefits while minimizing costs. In this sense, Nobel Prize-winning behavioral economist Richard Thaler (2000) explains, Homo Economicus—the subject of economic rationality—is assumed to be an incredibly sophisticated actor, a fast learner that never makes the same mistake twice, and a cerebral being whose decisions remain unaffected by emotions or the vicissitudes of everyday life. It is clear to see why Homo Economicus was so attractive for conceptualizing and then universalizing certain economic principles. But not without flaws. As Thaler argues, Homo Economicus is thoroughly abstracted, characterized in homogeneous ways that mix normative and descriptive elements, and assumes unrealistically that the environment in which economic actors operate is stubbornly static. Ironically, Homo Economicus manifests a form of self-delusion by economists who "base economic models exclusively on rational representative agents, while at the same time thinking that most of the people they interact with are at least occasionally bozos" (Thaler 2000, p. 136). Nonetheless, through the prism offered by the information deficit and economic rationality models, pursuing sustainable behavior appears as simply a matter of perfecting the informational environment to the point where sustainable behavior becomes irresistible. Alas, economic models do not always square with reality.

FAREWELL HOMO ECONOMICUS: WITHER INFORMATION DEFICIT MODEL

During the last three decades, our understanding of human behavior has become richer and much more nuanced than what the information deficit and the economic rationality models allowed. Current perspectives posit the relations between information and action as the dynamic outcome of several interacting intrinsic and extrinsic elements. Dan Lockton (2013), drawing on Herbert Simon's work on rationality and behavior, differentiates between cognitive and contextual factors. Cognitive factors include elements such as attitudes, motivations, cognitive biases, habits, valence and emotional salience, information processing, and knowledge acquisition. Contextual factors pertain to the physical and social environment, and include elements such as social practices and institutions, norms, values, and social imaginaries. Additionally, there are elements that seem to cut across the two groups. For instance, demographic characteristics such as age, level of education and level of income (Guerin et al. 2000), individual and group identity (Koger and Winter 2010; Wynne 1992), past and current social practices (Shove 2010a), and cultural values and predispositions (Crompton 2010; Kahan et al. 2011)—all seem to blend cognitive and contextual, intrinsic and extrinsic factors. Importantly, it is seldom the case that behavior change could be achieved by targeting only a single dimension: "Since different individuals face different impediments to behavior change and the impediments are often multiple, little happens until the right combination of intervention types is found" (Stern 2000, p. 419). Design, however, is situated precisely where context and cognition meet and is therefore well suited to influence behavior holistically (Lockton 2013, p. 40).

Homo Economicus may have withstood the dark ire of Nietzsche's philosophy and the cutting disenchantment of Freud's psychoanalysis, but its empirical discrediting accomplished by cognitive scientists, social psychologists, and behavioral economists, proved too much. The view of man as "omniscient calculator" (Lupia et al. 2000, p. 8), which dominated the European Enlightenment's self-image from Descartes to Kant and beyond, was no longer tenable. Since there was nothing lofty about reason, it could no longer be seen as the seat of human judgment. As cognitive scientists showed, reason is embodied and relies on our instincts, intuitions, emotions, and feelings (Damasio 1994, 2003; Lakoff and Johnson 1999). Body and mind, reason and emotion, are hopelessly entangled or, as Davidson (2000, p. 92) put it, "Cognition would be rudderless without the accompaniment of emotion, just as emotion would be primitive without the participation of cognition." Furthermore, embodied, sensorial, or unconscious responses are as crucial to our survival as our more calculative, logical, or conscious ones (Slovic 1987). Denying that fact is no more than fanciful self-deception. But there was more. As shown in Daniel Kahneman's Nobel Prize-winning research (alone, and with Amos Tversky), decision-making in difficult situations—when information is scarce and unreliable or when decisions need to be made fast—operationalizes a host of cognitive "heuristics." These are ingrained and intuitive mental strategies that allow people to "assess probabilities and predict values" with minimum effort (Tversky and Kahneman 1982, p. 4). While these economies of attention were honed over millennia to save human energy in situations fogged by uncertainty, in more complex situations, they introduce biases that may distort rather than reflect the real dimensions of the situation. Among these heuristics, Tversky and Kahneman note our tendency to evaluate probabilities based on the similarity of one case to others ("representativeness") or on the resemblance of a situation to easily retrieved, past occurrences ("availableness"), and our tendency to start making estimates based on a factual anchor while adjusting according to the specifics of the problem ("adjustment and anchoring"). The outcome of these thinking shorthands is that "when faced with a difficult question, we often answer an easier one instead, usually without noticing the substitution" (Kahneman 2011, p. 12).

Furthermore, our decision-making faculties appear to produce ossified, snowballing behavioral patterns that are "resistant to change because they influence the way that subsequent information is interpreted. New evidence appears reliable and informative if it is consistent with one's initial beliefs; contrary evidence tends to be dismissed as unreliable, erroneous, or unrepresentative" (Slovic 1987, p. 281).[6] Resistance to information that cannot be easily settled into existing cognitive molds, however, is not only contingent on intrinsic factors such as instincts or intuitions, but also on extrinsic factors such as societal values, norms, group identity, and social imaginaries (Ajzen 1991; Koger and Winter 2010; Krimsky and Plough 1988; Stern 2000; Taylor 2004; Wynne 1992). One need not look any further than the current state of climate change discourse and what some are calling "post-truth politics" to be reminded that the processing and rendering of information is influenced by the degree to which the information confirms or contrasts with people's values, ideologies, worldviews, and political orientation (Crompton and Kasser 2009; Hornsey et al. 2016; Nyhan and Reifler 2010; Tavris and Aronson 2007). We live in cognitive silos made by our ideological dispositions (see also Egan and Mullin (2017)).

Taken together, the insights of cognitive science, social psychology, behavioral economics, and other cognate disciplines attest to the degree to which human behavior is complex and situated: it depends on who is acting, where and when, and on the actor's values, identity, and social ties. Promoting sustainable behavior, it follows, is far more nuanced and far less deterministic than was assumed by the proponents of the information deficit model. To be clear, it is not that information is needless or useless, although sometimes information may indeed be inaccurate, irrelevant, or may contrast with strong preexisting worldviews and values to the extent that it becomes unbelievable. It is simply the case that information in and of itself is an insufficient condition for action. To trigger,

---

[6] This is sometimes referred to as "confirmation bias."

evoke, or even compel sustainable behavior, information must be combined with other, more direct behavioral prompts. Delivering such prompts is the aim of persuasive design.

Persuasive Design

The conceptual roots of persuasive design as a field of inquiry are often traced back to Richard Buchanan's articulation of design as rhetoric, and Madeleine Akrich's conceptualization of technical intentionality (Boks et al. 2015; Verbeek 2006; Zachrisson and Boks 2012). In "Declaration by design," Buchanan (1985) asserts the persuasiveness of design by relying on Aristotle's classic categories of rhetoric: *logos* or "design argument" refers to how design manipulates materials and processes in order to address specific needs; *ethos* or "design character" points to the ways designers express themselves, their values, ethics, and character; and *pathos* or "design emotion" indicates how designers make users feel by creating desirous, valuable, and fulfilling experiences. By substituting words with things, materials, and processes, Buchanan argues, design can be seen as "an art of thought directed to practical action through the persuasiveness of objects" (ibid., p. 7). While Akrich shares Buchanan's view of design as the redrawing of the relations between users, objects, and the world, she sees this process unfolding in less hermeneutic and more behavioral terms. In "The de-scription of technical objects," and under the influence of Latour's Actor-Network Theory, Akrich (1992) argues that technical objects have the capacity to prescribe or "delegate" certain behaviors to their users. They do so as material instantiations or "translations" of their designers' beliefs about the world into a set of features or functionalities. What she calls "scripts" articulate the range of permitted functions inscribed in a technical object as a means to effectively narrow down or fix the ways in which the object can be used. Seen this way, scripts determine how much control to leave in the hands of users, and which behaviors to delegate to the machine—manifesting "a specific geography of responsibilities, or more generally, of causes" (ibid., p. 207). Designed objects, Akrich argues, "not only lead to new arrangements of people and things," but also "generate and 'neutralize' new forms and orders of causality and, indeed, new forms of knowledge of the world" (ibid.).

What stood true for the photoelectric lighting kit and electricity meters with which Akrich illustrates her argument is even more pronounced in digital, interactive media. This is largely due to the latter's ubiquity—we turn to them for news, entertainment, commerce, and sociality, practically everywhere and all of the time[7]—but also because interactive media combine the features of both interpersonal and mass communication, potentially harnessing the persuasive capacities embodied in both (Fogg 2003; Oinas-Kukkonen and Harjumaa 2009). In this sense, interactive media convince, relate, associate, prompt, and compel users to act. They are, in other words, "objects with intent" (Rozendaal 2016).

Fogg (2003), a foundational figure in persuasive design, points out that digital technologies are involved in a multitude of persuasive activities, the study of which he terms "captology." For instance, shopping websites drive up sales by recommending products; educational websites and applications inculcate learning and skill development; healthcare technologies encourage healthy activities like cycling and the cessation of unhealthy activities like smoking; political websites urge users to participate in campaigns, and so forth. But most such efforts persuade *through* technology, and not *by* technology. In other words, they utilize or leverage new media as channels for delivering traditional modes of persuasion—much in the spirit of "digital rhetoric" (Losh 2009; Zappen 2005). While persuasion through technology takes advantage of the accessibility and ubiquity of modern devices, when it comes to persuasion by technology, it is interactivity that is the key feature. In Fogg's (2003) words, "As a general rule, persuasion techniques are most effective when they are interactive, when persuaders adjust their influence tactics as the situation evolves" (p. 6). We will see in the next chapter that the interactivity of new media is also essential to their capacity for "procedural rhetoric" (Bogost 2007), a mode of rhetoric that instead of using language or imagery deploys rule-based representations of the world. But at this time it should be noted that while customized responsivity allows new media to exceed the persuasive capacities of the traditional, mass media, other features allow them to exceed the capacities of interpersonal persuaders: new media are more persistent, allow greater anonymity, are able to store and use large amounts of relevant data, are able to combine various persuasive modalities (including content and form), can scale up to reach many users, and are ubiquitous (Fogg 2003, pp. 7–11).

Fogg (2003) notes that as agents of persuasion, computers (and indeed all new media) perform three discrete roles (what he calls the "functional

---

[7] As made evident by the range of reports from the Pew Institute (http://www.pewinternet.org) and the International Telecommunications Union (https://www.itu.int).

triad"). As *tools*, new media can persuade by making certain behaviors easier or harder, lead through processes to particular outcomes, provide timely information and feedback, and perform calculations that exemplify certain facts or realities. As *media* (symbolic and sensory), digital devices persuade by providing users with vicarious experiences, simulating situations that manifest particular cause and effect relations, or allow users to rehearse a desirable behavior while providing them with virtual feedback and rewards. Lastly, as *social actors*, new media persuade by providing socially significant cues to desired behavior. Such cues may include aesthetically or otherwise attractive, emotionally stirring, reciprocal, and rewarding interactions, or they may project authority. Whether they function as tools, media, or social actors, persuasive digital technologies share the basic rationale behind all persuasive media, that is, they seek "to cause a cognitive and/or an emotional change in the mental state of a user to transform the user's current state into another planned state and to cause a corresponding change in the user's behaviors" (Oinas-Kukkonen 2013, p. 1226). They may do so by *reinforcing* existing attitudes or behaviors, *changing* the way users respond to an existing issue or situation, or *shaping* new behavioral patterns in relation to new issues or situations. And they may do so in more or less apparent ways, while aiming for short-term compliance or long-term impacts (Oinas-Kukkonen 2013; Oinas-Kukkonen and Harjumaa 2009, p. 486; Tromp et al. 2011). In what follows, I discuss several persuasive interactions. My analysis foregrounds the extent to which they rely on infocentric strategies, and the degree to which they maintain user autonomy.

RESPONSIVE, ACTIONABLE INFORMATION

Boks et al. (2015) note that early work in design for sustainable behavior focused on the material lifecycle of artifacts: ways to improve disassembly and recycling, and to extend the lifespan of products in order to reduce the need to reuse or recycle them. Around the mid-2000s, designers started shifting their attention toward products-in-use and, more importantly for the present context, began addressing user-artifact interactions more explicitly. In this context, the first group of designs discussed here aims to raise awareness of unsustainable consumptive behaviors under the assumption that this will increase intensions to act and, ultimately, reshape behavior in more sustainable ways. The role of information is central, but unlike mass media informational campaigns, the provision of information here is both responsive and coupled to achievable goals. The information, in other words, is not only meant to be actionable (i.e., includes elements that make acting on it both possible and appealing), but is dynamic enough to reflect user actions back to them. The behavioral loop between information and action can therefore be tightened, potentially closing the attitude-behavior gap.

Take, for example, Coralog, an ambient display widget that visualizes the computer's consumption of electricity as the health state of a coral reef (Kim et al. 2010). The widget takes as its input energy consumption trends, or more accurately, the relation between the computer's uptime (when it consumes energy) and its idle time (when it consumes energy but does not perform calculations). If the relation between idle time and total uptime is better than it was before (i.e., indicating a positive trend), the coral reef appears healthier—more colorful and with more abundant coral and fish. Inversely, a negative trend will cause the reef to blanch and lose biomass (see Image 2.2).

Coralog's visualization not only brings to light a hidden aspect of energy waste (computation idleness), but it makes iconic use of the natural environment as a means to make energy consumption less abstract. Using the reef as a visual metaphor helps make tangible the connection between domestic energy consumption and its environmental effects. Additionally, Coralog targets a specific aspect of electricity use (a micro-activity) and therefore provides users with a clear call to action. This was reflected in the widget's impact. Although the sample size was quite small (17 users), in two weeks of field testing, widget users reported having an increased interest in their computer electricity usage, with 30% of them reporting that they actually tried to change their behavior by, for instance, using sleep mode or turning off their computers more often—a claim supported by actual log data retrieved directly from users' computers (Kim et al. 2010, p. 108). So while it is entirely reasonable to be skeptical about the degree

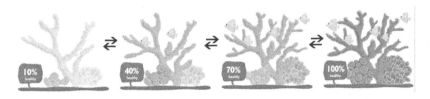

Image 2.2   Coralog's ambient display. (Image by Kim et al. 2010)

to which Coralog's intervention would last, that is, whether users would maintain or even deepen their response to the bleaching reef over time, it is evident that the widget provided at least some motivation for users to dig deeper into their consumption habits and attempt to change them.

As Coralog shows, targeting particular domains or micro-activities can be an effective behavior change strategy. A similar approach motivates eco-driving applications. The one described by Gabrielli and Maimone (2014), for instance, is capable of using the mobile phone's Bluetooth connection to retrieve information from the car's engine, including acceleration, deceleration, average fuel consumption, speed, and engine temperature, which it then displays on an "eco-dashboard." Drivers receive a dynamic eco-driving score, and can compare their current driving performance to past journeys. The app's more immediate behavioral incentives are the context-aware visual and audial cues it generates algorithmically. The system thus encourages drivers to drive differently by providing them with "just-in-time" audiovisual advice based on their driving profile and the car's current state, prompting drivers to reduce the number and frequency of accelerations and breaks, close windows when driving in high speed, use the appropriate gear when driving uphill, and so forth. Simpler eco-drive applications such as greenMeter rely on the mobile phone's own GPS and accelerometer instead of querying the car's engine. They have been on the market for over a decade and are generally viewed positively, yet evidence of their effectiveness in terms of both fuel conservation and CO_2 emissions reduction is rather mixed.[8]

Coralog and greenMeter can be seen as forms of "eco-visualization" (Pierce et al. 2008) or "eco-feedback" techniques. These aim "to inform users clearly about what they are doing and to facilitate consumers to make environmentally and socially responsible decisions through offering real-time feedback" (Bhamra et al. 2011, p. 431). Eco-feedback techniques may address a range of consumptive behaviors including water usage, energy and fuel consumption, choices over waste disposal, recycling, and

---

[8] Hiraoka et al. (2009) report positive outcomes, but their test was done with simulators. Lee et al. (2010) report very minor changes in driving behavior based on both online surveys and actual user tests data. The makers of greenMeter, however, make the following statement on their website: "Based on a conservative estimate of worldwide use in its first year, green-Meter has *likely* saved about 2 million gallons of gas, reduced fuel expenses by over $5 million, dropped oil consumption by nearly 20,000 barrels, and prevented almost 47,000 tons of CO_2 from entering the atmosphere" (emphasis added; http://hunter.pairsite.com/greenmeter (last accessed Mar. 18, 2018)).

transportation (Froehlich et al. 2010). By targeting micro-activities, as mentioned above, eco-feedback techniques draw clearer, more personal links between information and possible courses of action, and therefore may be successful in shaping intentions to act and actual behavior. In this sense, while eco-feedback techniques remain committed to the centrality of information for reforming sustainable behavior, because they are "tailored to the intentions, capabilities and expectations of the individual" (Wilson et al. 2015, p. 188), they no longer exhibit the one-size-fits-all logic that undergirded the information deficit model. Consequently, the success of eco-feedback techniques relies on their designers' ability to understand both the contexts of behavior and the ways in which communicating (or feeding back) the information may be most effective. This may be achieved by paying attention to the information's accuracy, granularity, and comprehensiveness, the frequency and duration of the feedback, and the degree to which the feedback becomes actionable in terms of how it fits, confirms, or challenges the user's existing normative, valuative, and cognitive frames (Wilson et al. 2013). But designing effective eco-feedback mechanisms also requires developing an informed expectation of where the most effective, "teachable moment" may lie, and how much control users actually have over the behavior domain. Inhabitants of residential homes, dormitories, public buildings, and offices, for instance, enjoy different degrees of power over climate control and would therefore require different eco-feedback solutions (Pierce et al. 2008).

Given the importance of context to the effectiveness of persuasive design, it is worth describing with some detail how eco-feedback mechanisms may be customized. In the case study described by Wilson et al. (2013), the customization of the eco-feedback mechanism included several phases. First, the designers selected for their intervention seven social housing tenements in the south of Wales (United Kingdom). These featured different variations in terms of household composition, the age and form of the houses, the heating systems they used, and the way they paid for energy. Second, after investigating the settings through a combination of interviews and observations, the designers generated "areas of opportunity," that is, everyday domains in which the design intervention may be most effective. In this case, the chosen "area of opportunity" pertained to contrasting demands for heating and fresh air. Next, the designers created a design brief (a clear definition of the problem) and articulated the "solution space" by developing relevant design concepts and selecting the most appropriate design direction. Lastly, designers created and deployed the eco-feedback solution: a system that monitored the state of both the heating system (by measuring the surface temperature of radiators) and the window's position (indicating whether windows were open or closed). Interactive feedback was provided in the form of light and sound indicators, and reflected back to users the degree to which their behavior was deemed "wasteful" (when the radiator was active while the window was open). In order to evaluate the appropriateness of their solution, designers conducted focus groups with ten households, and user trials with two of them. They then developed a pre-intervention qualitative baseline in relation to the two specific households in which the eco-feedback devices were to be installed, and let participants themselves select the specific locations for installation.

Despite the meticulous preparation and evidence that the use of the eco-feedback device had deepened participants' understanding of both their heating systems and their own behavior patterns, the intervention described above resulted in no discernable behavior impacts. Could a better designed intervention yield more conclusive behavior change? Is the designated behavior even susceptible to feedback-type interventions? Is it reasonable to expect that participants will adjust their behavior in accordance to a series of lights and clicks or, inversely, be willing to invest mental energy in continuously evaluating the appropriateness of their behavior? It remains unclear.[9] However, Wilson et al. (2015) note that it is not uncommon for eco-feedback interventions to have only a modest behavioral impact since they are prone to several complications: the information may prove partial and irrelevant or may lack motivational power; the intervention may target one user group but may involve others and therefore prove insufficient for those unintended users; the feedback mechanism may fail to attract sufficient attention and fade into the background; and feedback devices may cause or amplify various "rebound effects" with consequences inverse to the intervention's intentions. Setting concerns over poor usability aside (but see, for instance, Peffer et al. 2011), the problem could be that eco-feedback interventions are too removed from the everyday practices they seek to reshape. Such concerns animate social practice theory and prompt Yolande Strengers (2011) to argue that treating household members as "micro resource managers" is blind to the various ways

---

[9] To some extent, these questions relate to larger concerns about the long-term sustainability of eco-feedback interventions (see, for instance, Barreto et al. 2013; Snow et al. 2013).

in which resource consumption is entangled with past experiences, a variety of everyday activities, interactions with family and friends, and the presence of other relevant technologies. Reductions in resource consumption, she adds, require a deep understanding of the entire network of social practices in which the particular consumptive activity takes place—a shift of focus from individual agents to the social routines in which they partake (see also Strengers and Maller (2015, p. 3)).

Clearly, context matters. But perhaps the reason for the modest success of eco-feedback interventions is far more trivial. This is because awareness raising, much like other infocentric interventions, maintains *by design* a large degree of user autonomy over the media, and given such autonomy, users may, well, act autonomously. As Zachrisson and Boks (2012, p. 58) point out, for informational strategies to work, "the user has to take in the information, and be willing to change the behaviour. This implies that the user should have a positive attitude or be motivated to perform the intended behaviour,"[10] which, I believe, gives way to a veritable chicken and egg conundrum: information is expected to generate motivation, yet without motivation it is unlikely that the information will be effective to begin with. In other words, infocentric persuasion relies on deliberate, cognitive evaluation and therefore remains largely within the space carved by the information deficit model. Participants in the Wales study may have been generally motivated to take part in the experiment but less motivated to follow the behavior scripts suggested by the eco-feedback mechanism.

The eco-feedback techniques discussed above attempt to influence behavior by coupling information with action. But while they emphasize the informational aspect of the coupling, "serious games" shift the emphasis to actionability.[11] Of course, some serious games merely "gamify" responsive information delivery, adding playfulness to what are essentially eco-feedback interventions (see, for instance, Gamberini et al. 2012; Geelen et al. 2012), but other games attempt to influence users not by providing them information just-in-time but by their very interactive structures, "seducing" (Tromp et al. 2011) them to pursue certain behaviors by deploying in-game rules and procedures. Take, for example,

---

[10] Yang et al. (2014) make the same point: "considerable motivation and engagement on the part of consumers is required for eco-feedback to lead to behavior changes" (p. 824).

[11] Serious games can be defined in different ways, but in essence they attempt to transfer in-game behavior to real-world issues and contexts—"from the game world into the material world" (Bogost 2007, p. 47).

Carbon Chaos, a mobile game designed by students at the Centre for Digital Media in Vancouver.[12] The objective of the game is to transport quickly passengers from one side of a city block to their destination on the other edge of the screen. The user may choose one of three modes of transportation: bicycles carry only one person, cars carry up to three persons, and buses carry up to ten persons. While bicycles do not emit CO_2, cars and busses do, leaving behind a thick cloudy trail that, until it dissipates, halts other cars and buses. (Bicycles, on the other hand, and in clear violation of physiology, remain immune to the halting effects of the CO_2 trail). Once passengers are dropped successfully at their destination, the player receives a score and their "Greenroof Meter" is charged. If the Meter is full, the player can tap on a destination and release a "shockwave" that clears all carbon clouds from the map.

Carbon Chaos's strategy for persuading players to use less carbon-intensive transit options is premised in what Fogg (2003) calls "tunneling": leading users through tightly controlled, predetermined interactive structures. In Fogg's words,

> Tunneling technologies can be quite effective. For users, tunneling makes it easier to go through a process. For designers, tunneling controls what the user experiences – the content, possible pathways, and the nature of the activities. In essence, the user becomes a captive audience. (p. 36)

Tunneling effects are evident in both *Carbon Chaos*'s outcome and the interactivity it affords its players. The game's singular objective, to score more points and "level- up," colors the gameplay experience. The kind of interactive agency the player enjoys remains largely subordinated to a principle of efficiency: the player cannot change the scene, add or remove means of transportation, or manipulate the rules governing emissions and scoring. What they can do is play the game better, that is, score more points by being more "sustainable." Under these conditions, it is unclear whether playing the game successfully conjures a deeper understanding of the infrastructure or policy implications of transit-related choices, or whether it merely teaches players how to play the game with more dexterity.[13] The game's potential influence on "real-world" behavior, however,

---

[12] See more about the game here: https://thecdm.ca/projects/archives/carbon-chaos-2010 (last accessed Mar. 18, 2018).

[13] It appears the team developing the game "didn't have the resources to determine if the game actually increases environmental awareness," admitted the professor who guided the process of development (cited in Lavender 2010, July 18).

does not rely on the depth or accuracy of the information it offers players. Instead, it relies on the fact that there is only one way to play the game, with very little room for exploration, experimentation, and reflection. To play the game "right"—to score high and win—the player needs to repeatedly enact the underlying message that bicycles and public transportation are better environmental choices than cars. Players may be hard pressed to find other meanings in the game, but it is precisely this one-dimensionality that reflects the game's designers' intent to persuade.

LESS SEDUCTION, MORE PERSUASION

Whether they opt to prompt users to act with customizable, just-in-time information as eco-feedback mechanisms do, or seduce them with playful tunneling as *Carbon Chaos* does, the media discussed above maintain a large degree of user autonomy. They may be more or less explicit about their goals, but they remain content with providing users with actionable information. Some design strategies, however, choose a more direct approach and attempt to push or "nudge" (Thaler and Sunstein 2008) users to change their behavior regardless of whether users understand the sources or implications of their behaviors. Zachrisson and Boks (2012) state the essence of such interventions:

> the user is still in charge, but the product takes more control by making the desired behaviour easiest or most intuitive. These strategies can be assumed not only to be effective on users with a positive attitude but also on users who do not have a particular attitude. As the desired behaviour is easiest, this is what the user can be expected to do, as long as no effort is made to behave in another way. (p. 58)

What Zachrisson and Boks are describing is equivalent to what Bhamra et al. (2011) call "eco-spur" and "eco-steer": the former aims to inspire and incentivize sustainable behavior, while the latter tries to prescribe sustainable behavior through affordances and constraints embedded in the product itself.

We can get a sense of how eco-spur techniques work by looking at Flower Lamp (see Image 2.3). The lamp was part of Static!, a design project by the Swedish Interaction Design Institute. The project, which ran

Image 2.3 Closed and open Static! Flower Lamp. (Image by Interactive Institute, Sweden)

between 2004 and 2006, sought to increase awareness of everyday domestic energy consumption while experimenting with the materiality, expressivity, and aesthetic qualities of energy itself (Backlund et al. 2006; Mazé 2010). Through a series of energy-aware domestic objects such as a curtain, radiator, lamps, a door, tiles, and an extension cord, the project offered thought-provoking ways to render the hidden omnipresence of energy visible and therefore actionable. Although the lamp is not digital, the manner in which its responsive aesthetic anchor its persuasive effect is instructive. The lamp provides a physical manifestation of the household's electricity use, taking as its input the household's energy consumption trend, that is, the degree to which the household has been increasing or decreasing its overall energy consumption. A positive trend causes the lamp to open its form (or "bloom"), while a negative trend causes the lamp to close down by bunching the petals. The lamp, in this mode, embodies a specific behavioral script: it incentivizes more frugal energy consumption practices by its very shape. If users want to enjoy the full beauty of the lamp, and a larger illuminated area, they need to reduce their energy consumption. Function and aesthetics combine here to produce a behavioral nudge.

Physical incentives such as the ones illustrated by Flower Lamp are both intuitive and difficult to resist, but users could also be nudged to act by making the activity socially visible, generating positive reinforcement (increasing social status) or negative reinforcement (dealing with mockery or ridicule). Take, for example, StepGreen, a plugin designed to encourage pro-environmental behavior by making it visible on social networks. The plugin provides users with various suggestions for actions that carry tangible environmental and financial benefits (or "savings"). It then allows them to commit to action, follow-up on their actions, and share their progress via social networks such as Facebook (Mankoff et al. 2010). The incentivizing mechanism deployed by StepGreen is threefold: first, the actions are presented in ways that are both accessible and easy to accomplish; second, the user's progress is visualized in a way that simplifies keeping track of both accomplishments and incomplete actions; and third, the user's performance is made shareable and comparable to others, thus inviting users to commit and accomplish more actions as a means to increase social status. Whereas StepGreen relies on user self-reporting, BinCam (briefly mentioned at the top of this chapter), provides "objective" evidence of pro-environmental behavior. The device includes a camera installed in the household's garbage bin and an app that shares the images online. Every time a member of the household throws something in the bin, the camera snaps an image. The image is then uploaded and analyzed via Amazon's Mechanical Turk crowd-sourcing service for the total number of items disposed, the number of recyclable items, and the number of food items. This data is appended to the image, which is then shared to other BinCam members and can be shared even more widely via Facebook (Thieme et al. 2012). A gamification mechanism ("BinLeague") is used to spur competition between households, amplifying the device's capacity to promote "normative social influences" and thus support behavior change more actively (ibid., p. 2341).

BinCam's use of social visibility proved to be only partially successful. "Participants stated that pictures of other people's rubbish were not of much interest to them and focused mainly on those activities of their own bin and household" (ibid., p. 2342). Neither were participants particularly enthusiastic about the competition. This, however, was not the case with Wattsup (not to be confused with the communication app, WhatsApp), an app that connects the Wattson Home Energy Monitor to Facebook (Foster et al. 2010). The app combines a list of commitments (such as the one used by StepGreen) with a ranking mechanism (such as the one used by BinCam), making every participating household's energy use both visible and competitive. When comparing the energy consumption of those using Wattson socially (with Wattsup) with those who did not, the results were encouraging: "energy consumption was significantly lower when using the socially enabled application" (ibid., p. 183), and participants seemed to take pleasure in the competitive elements (ibid., p. 184). This led the designers to conclude that "Competitive carbon counting appears to be both more enjoyable and more effective than individual monitoring" (ibid., p. 185). As the three examples discussed above illustrate, social incentives may be persuasive without the need to develop a comprehensive awareness of the environmental implications of current or desired behaviors, and without fully customizing the intervention to fit particular conditions or settings. One may imagine users engaging with StepGreen, BinCam, or Wattsup not because they are motivated by environmental concerns but because they seek visibility and higher social status.

A TECHNOLOGY OF SUSTAINABLE BEHAVIOR (OR, A SUSTAINABILITY SKINNER BOX)

Whether they attempt to seduce the user or nudge them to act, the examples discussed so far still leave a significant degree of autonomy to the user. They raise awareness of the consequences of inaction, aim to develop intentions to act, or prompt users to take action, but they do not force a particular behavior. Erratic Radio, which, like Flower Lamp, was part of the Static! project is different. The radio was designed to respond to the presence of other electrical appliances. Its regular receiver is controlled by an additional one that is tuned to frequencies around 50 Hz (the range of frequencies emitted by most appliances). When the radio senses an increase in the electrical field surrounding it, indicating the presence of other working appliances, it loses its tuning. Unless listeners shut down nearby appliances, the radio becomes an unlistenable, unusable static generator. Unlike the media discussed above, Erratic Radio forces users to act, and in this sense it puts concrete behavior ahead of attitudes or intentions to act (see also Hassenzahl and Laschke 2015). Instead of assuming that changing attitudes would lead to concrete actions (consistent with Fishbein and Ajzen's theory of reasoned action or with Ajzen's theory of planned action), Erratic Radio flips the script, manifesting a kind of sustainability Skinner box.

A Skinner box is an experimental contraption used for operand conditioning (training subjects to respond to stimuli in a specific way). It embodies the principles of what American behavioral psychologist B.F. Skinner (1971) called a "technology of behavior": a mechanism of behavioral manipulation that operates with scientific precision, consistency, and reliability. A technology of behavior projects scientific certainty into the realm of human behavior:

> We could solve our problems quickly enough if we could adjust the growth of the world's population as precisely as we adjust the course of a spaceship, or improve agriculture and industry with some of the confidence with which we accelerate high-energy particles, or move towards a peaceful world with something like the steady progress with which physics has approached absolute zero (even though both remain presumably out of reach). (ibid., p. 5)

While the goals sought by the technology of behavior may be more or less socially beneficial—as the brief citation above shows, Skinner's own aims were quite benevolent, even utopian (Rutherford 2003)—the means by which the technology seeks those goals replaces cognitive manipulation with environmental triggers. This is consistent with Skinner's belief that it is environmental conditioning and not man's free will that shapes human behavior. Much to the chagrin of his critics (see, for instance, Strike 1975), Skinner insisted that human autonomy was nothing but an illusion: "Autonomous man serves to explain only the things we are not yet able to explain in other ways. His existence depends upon our ignorance, and he naturally loses status as we come to know more about behavior" (Skinner 1971, p. 14). If humans are never really autonomous, Skinner adds, we may as well learn how to engineer their behavior in scientific, reliable ways that may benefit society.

With the dissolution of both Homo Economicus and "autonomous man," and given the scope of environmental issues and the kind of urgency they raise, pursuing a technology of sustainable behavior sounds quite appealing. Who would reject designed environments that prohibit littering, houses that enforce energy savings, or public transit systems that prove irresistible? But considering what we know today about the complexity of human behavior, the kind of certainty Skinner sought could only be attained by applying a considerable degree of coercion that many would find inacceptable—or at least distasteful. Nonetheless, given interests in subliminal priming of pro-environmental behavior (Ruijten et al. 2011), and increasing efforts to embed behavioral cues into the built and virtual environment, the very possibility of realizing "sustainability by stealth" (Lilley et al. 2005) raises important ethical concerns. As displayed by the examples presented above, different persuasive interactions embody varying degrees of paternalism, affording users with different degrees of control over the interaction and its outcomes. They also imply a different measure of transparency (the degree to which the user is aware of the intervention) and voluntariness (the degree to which the user initiated the intervention), although even users who are interested in modifying their behavior may not be aware of the full extent or implications of the designed intervention. Consequently, as the interaction becomes more persuasive, the user enjoys a diminishing degree of autonomy, understood here as the capacity to identify, modify, and resist the kind of behavior prescribed by the technology. Persuasion can easily turn into coercion, or worse, into manipulation (Smids 2012). Persuasive design can become "evil by design" (Nodder 2013).

Unsurprisingly, ethical concerns over persuasive technology bear the traces of previous concerns about the persuasive powers of the mass media (see, e.g., Dawson 2003; Ewen 2001; Packard 1957). While questions about the values implicit in design or in computation are not new (see, for instance, Friedman 1996; Friedman and Nissenbaum 1996), concerns about persuasive technologies emerged largely in response to Fogg's "captology" in the late 1990s (as described above). In a paper published just before the turn of the millennium, Berdichevsky and Neuenschwander (1999) ask:

> What if home financial planning software persuaded its users to invest in the stock market? And what if the market then crashed, leaving the users in financial ruin? Or, more subtly, what if the makers of the software arranged with certain companies to 'push' their particular stocks? Would such designs differ in a morally relevant way from stockbrokers who encourage their clients to buy the stocks that earn them bonus commissions? (pp. 51–2)

Their answer foregrounds the same principles noted above as the very strengths of persuasive technologies over traditional means of persuasion, namely, the former's capacity to interact dynamically with users and by that provide "The appearance of control" (ibid., p. 52) while effectively undermining autonomy. Nonetheless, Berdichevsky and Neuenschwander write, the computational device alone cannot be blamed for the consequences of persuasion. "Rather, responsibility for the computerized machine's built-in motivations, methods, and outcomes falls squarely on its creators and purchasers" (ibid., p. 54). This leads them to suggest eight principles for the ethical design of persuasive technologies, capped by a kind of Rawlsian "golden rule" of persuasion: "The creators of a persuasive technology should never seek to persuade anyone of something they themselves would not consent to be persuaded of" (ibid., p. 59). As a rule of thumb, this suggestion may have universal appeal, but it leaves unaddressed the fact that people may have very different ideas about what they may or may not be willing to be persuaded of. It also assumes that people actually know when they are the subjects of persuasive design. But do they? *Can* they?

To answer these questions, users (and researchers) need to first be able to recognize persuasive technologies *as such*. This is the essence of Brey's (2000) proposal for "disclosive computer ethics": "disclosing and evaluating embedded normativity in computer systems, applications and practices" (p. 11). It is also the rationale behind more institutional or bureaucratically oriented suggestions that authorities protect users from the nefarious effects of persuasive technologies by appropriately labelling those technologies, or by compelling designers of persuasive interactions to be accredited or certified (de Oliveira and Carrascal 2014). Whether or not such measures will be enacted remains to be seen. But even though it is clearly evident that designers of persuasive interactions are indeed aware of the ethical implications of their work (for a few examples, see Davis 2009; Karppinen and Oinas-Kukkonen 2013; Lilley and Wilson 2013; Pettersen and Boks 2008; Timmer et al. 2015), a fundamental conundrum persists: unless we are dealing with self-generated goals or willing to dilute the potential impacts of the intervention by revealing it upfront, persuasive interactions will always reflect a kind of paternalistic spirit. For at least some, persuasive design remains uncomfortably close to social engineering.

BEYOND BEHAVIOR CHANGE

The image of sustainability that emerges from the design and use of persuasive interactions is one of a balancing act—an effort to regain a delicate, uneasy equilibrium between the material needs of humans and those of nonhumans. As long as humans continue to extract, consume, and pollute with reckless abandon, the planet's ecosystems and inhabitants are destined to suffer. More than a few species will go extinct. Many already have.[14] As David Wallace-Wells (2017, July 9) warns in a much discussed article in *New York Magazine*, "absent a significant adjustment to how billions of humans conduct their lives, parts of the Earth will likely become close to uninhabitable, and other parts horrifically inhospitable, as soon as the end of this century." Persuasive interactions aim to halt this trajectory and right the planetary balance by rerouting the way we conduct our lives.

Setting aside ethical concerns, upon encountering the range of persuasive interactions discussed here, the reader may wonder about the degree to which they are effective. This is, of course, a legitimate question, and one that has been, and is still raised in regard to other media (Gauntlett 2005) and to other persuasive domains such as advertising (Kim 1992; Samuel 2010; Vaughn 1980). The answer, however, is not straightforward. It may be because the field is relatively young and still in need of discovering both its potentials and its limitations, but it may also be the outcome of the way designers of persuasive interactions for sustainability frame, evaluate, and report their interventions. As illustrated by the comprehensive reviews of Brynjarsdóttir et al. (2012) and Hamari et al. (2014), the results of persuasive interactions are often based on fairly small sample sizes,[15] are not always subjected to thorough evaluation by recognized measurements or scales, and rely too heavily on user self-reporting that may be inaccurate (see also Mankoff et al. 2010, p. 110). Some of the interventions discussed above (Flower Lamp and Erratic Radio in particular) are more artistic in nature and therefore make no claims of measurable effectiveness, while others may not be mature enough to advance beyond early stage usability tests. Furthermore, some of the intended or anticipated effects may take a while to unfold, may be cumulative in nature, or may be simply too nuanced to detect. Overall, Niedderer et al. (2016, p. 76) argue, "there is an urgent need for more explicit information and debate about the aims and benefits of DfBC [design for behavior change]," and a need to strengthen "the evidence base on the impact of DfBC" by developing "clear evaluation metrics for DfBC." This, of course, equally applies to the design of persuasive interactions for sustainability.

With that said, perhaps the problem is not so much that persuasive interactions for sustainability lack "clear evaluation metrics" but that they

---

[14] See, for instance, the "Living Planet Index" by the World Wide Fund for Nature (WWF) (McRae et al. 2016).
[15] Mostly in the range of 5–15 participants according to Brynjarsdóttir et al. (2012, p. 949), a median of 26 according to Hamari et al. (2014, p. 123).

narrow the meaning of sustainability to such an extent that their results seem meager, negligible, or ungeneralizable (Brynjarsdóttir et al. 2012). Do Coralog users extend the lessons taught by the widget to other domains of energy use? Do players of *Carbon Chaos* understand the policy decisions that underpin the game's rules, and, as consequence, are they more likely to support investment in public transit infrastructure? Would greenMeter users be more likely to engage in other energy conservative behaviors? After the 5-week experiment ended, did users of BinCam continue to recycle "more and better" (Thieme et al. 2012, p. 2343)? Answering such questions requires designers to not only expand the depth and scope of the ways in which they evaluate the persuasiveness of the interactive media they create, but also to rethink the way they understand sustainability. As made clear by the examples discussed above, the overwhelming majority of persuasive interactions target individual, consumptive behaviors. But how effective can persuasive interactions be if they understand both human behavior and the conditions for its transformation exclusively through cognitivist or behaviorist prisms? As argued by scholars and designers working from a social practice perspective (Shove 2010b; Shove and Spurling 2013; Strengers and Maller 2015), and briefly discussed above in the context of eco-feedback interventions, any single individual behavior is always implicated with a matrix of collective concerns, practices, and experiences, and could therefore be understood as merely one "moment" in a larger continuum of unfolding social practices (Warde, cited in Strengers and Maller 2015, p. 2). Reducing human behavior to a series of more or less informed, and more or less motivated individual choices over consumption, in disregard of "the cultural, material and economic structuring of consumption" (Shove and Spurling 2013, p. 3), and with indifference to how the latter themselves change through social practices, risks producing partial solutions that are neither sustainable nor scalable (Brynjarsdóttir et al. 2012; DiSalvo et al. 2010; Kuijer and Bakker 2015).[16] Since cognition and context always work in tandem, behavioral interventions must account for both. What counts as context, however, may vary from immediate settings to larger social, cultural, and political structures—from micro- to macro-contexts.

On the subject of macro-contexts, persuasive design's explicit placing of the responsibility for sustainable change on individuals seems oblivious

---

[16] The exchange between Shove (2010a, 2011) and Whitmarsh et al. (2011) is particularly relevant here.

to the fact that the individual's capacity to act is conditioned by a host of social, economic, and political institutions. Many sustainable behaviors depend on public utilities and infrastructure—on what Lewin (cited in Lockton 2013, p. 41) calls "channel factors": sustainable mobility choices depend on the availability of efficient and convenient public transit; sustainable urban lifestyles are enabled by denser, mixed-use neighborhoods which, in turn, necessitate appropriate zoning regulations; reducing the household's material waste and increasing recycling behaviors depend on the availability of appropriate bins, but could be even more effective if grocery stores used less packaging. Other sustainable behaviors rely on economic or market incentives. The conditions for scaling up sustainable behavior, it follows, can largely be found in the domain of law, regulation, policymaking, and governance. Sustainable behavior, in other words, is conditioned by politics (Brulle 2010; Hulme 2009; Klein 2014; Lövbrand et al. 2015). From this perspective, persuasive design features a mismatch between "the scales of action and the scales of effects," as Dourish (2010, p. 1) aptly argues. A more effective design strategy, he adds, will help translate individual, technically-mediated experiences into collective action aimed at political domains:

> rather than using technology to provoke reflection on environmental impact of individual actions, we might use it instead to show how particular actions or concerns link one into a broader coalition of concerned citizens, social groups, and organizations.... By focusing not on connecting people *to* their actions and their consequences, but on connecting people *through* their actions and their consequences, we can approach persuasive technologies as ones whose intent is to persuade people of the effectiveness of collective action and of their own positions within those collectives. As an approach to the use of interactive technologies and environmentalism, it attempts to move from fostering environmental consumers to shaping environmental movements. (ibid., p. 7; emphasis in origin)

What Dourish calls "scale-making" points to the limitations of both the goals and the methods of persuasive design. In terms of goals, it speaks to the elephant in the room, that is, to the fact that transforming individual lifestyles to entirely sustainable paths will likely come short of the kind of results we need in order to halt global climate change and environmental degradation. In terms of methods, it articulates the need for designing micro-interventions that shift focus from individual to collective behavior, reversing the kind of individualization that undermines the scaling up of persuasive design.[17] "Sustainability does not begin with the individual," argue Hazas et al. (2012). Neither should it end there.

From the perspective of social practice theory, other forms of persuasive design focus too narrowly on the cognitive dimensions of individual behavior without paying sufficient attention to the network of practices, relations, and experiences from which every behavior emerges. From a more political or political-economic perspective, persuasive design seems to target the symptoms and not the root causes of unsustainable behavior, and should therefore seek ways to address collective behavior and the social and political infrastructure that conditions it. While persuasive interactions may help shift certain micro-activities and provide a useful vocabulary for considering the individual and personal dimensions of sustainability, on their own they seem woefully inadequate to affect large-scale changes. Picking up all the litter in Cornwall or even in the entire United Kingdom will hardly begin to make a dent in the planet's GHG calculus. Impacting large-scale phenomena, as discussed in the next chapter, requires a more holistic, systemic approach.

BIBLIOGRAPHY

Ajzen, I. (1991). The Theory of Planned Behavior. *Organizational Behavior and Human Decision Processes, 50*, 179–211.

Akrich, M. (1992). The Description of Technical Objects. In W. E. Bijker & J. Law (Eds.), *Shaping Technology/Building Society: Studies in Sociotechnical Change* (pp. 205–224). Cambridge, MA: MIT Press.

Backlund, S., Gyllenswärd, M., Gustafsson, A., Ilstedt Hjelm, S., Mazé, R., & Redström, R. (2006). *STATIC! The Aesthetics of Energy in Everyday Things.* Paper Presented at the Design Research Society International Conference, Lisbon.

Bamberg, S., & Möser, G. (2007). Twenty Years After Hines, Hungerford, and Tomera: A New Meta-Analysis of Psycho-social Determinants of Pro-environmental Behaviour. *Journal of Environmental Psychology, 27,* 14–25.

Bandura, A. (1994). Self-Efficacy. In V. S. Ramachaudran (Ed.), *Encyclopedia of Human Behavior* (Vol. 4, pp. 71–81). New York: Academic.

Barreto, M., Karapanos, E., & Nunes, N. (2013). Why Don't Families Get Along with Eco-feedback Technologies?: A Longitudinal Inquiry. In *Proceedings of CHItaly '13 (Article No. 16)*. New York: ACM.

---

[17] A few examples of work in SHCI that is already moving in this direction are Ganglbauer et al. (2013), Knowles et al. (2014), Prost et al. (2014).

Berdichevsky, D., & Neuenschwander, E. (1999). Toward an Ethics of Persuasive Technology. *Communications of the ACM, 42*(5), 51–58.

Bhamra, T., Lilley, D., & Tang, T. (2011). Design for Sustainable Behaviour: Using Products to Change Consumer Behaviour. *The Design Journal, 14*(4), 427–445.

Bickman, L. (1972). Environmental Attitudes and Actions. *The Journal of Social Psychology, 87*(2), 323–324.

Bogost, I. (2007). *Persuasive Games: The Expressive Power of Videogames.* Cambridge, MA: MIT Press.

Boks, C., Lilley, D., & Pettersen, I. N. (2015). The Future of Design for Sustainable Behaviour, Revisited. Paper Presented at the 9th EcoDesign International Symposium on Environmentally Conscious Design and Inverse Manufacturing, Tokyo.

Brey, P. (2000). Disclosive Computer Ethics. *Computers and Society, 30*(4), 10–16.

Brulle, R. J. (2010). From Environmental Campaigns to Advancing the Public Dialog: Environmental Communication for Civic Engagement. *Environmental Communication, 4*(1), 82–98.

Brynjarsdóttir, H., Håkansson, M., Pierce, J., Baumer, E. P. S., DiSalvo, C., & Sengers, P. (2012). Sustainably Unpersuaded: How Persuasion Narrows Our Vision of Sustainability. In *Proceedings of CHI '12* (pp. 947–956). New York: ACM.

Bucchi, M. (2008). Of Deficits, Deviations and Dialogues: Theories of Public Communication of Science. In M. Bucchi & B. Trench (Eds.), *Handbook of Public Communication of Science and Technology* (pp. 57–76). London/New York: Routledge.

Buchanan, R. (1985). Declaration by Design: Rhetoric, Argument, and Demonstration in Design Practice. *Design Issues, 2*(1), 4–22.

Burgess, J., Harrison, C., & Filius, P. (1998). Environmental Communication and the Cultural Politics of Environmental Citizenship. *Environment and Planning A, 30*, 1445–1460.

Crompton, T. (2010). *Common Cause: The Case for Working with Our Cultural Values.* Surrey: WWF-UK.

Crompton, T., & Kasser, T. (2009). *Meeting Environmental Challenges: The Role of Human Identity.* Surrey: WWF-UK.

Damasio, A. R. (1994). *Descartes' Error: Emotion, Reason, and the Human Brain.* New York: Putnam.

Damasio, A. R. (2003). *Looking for Spinoza: Joy, Sorrow, and the Feeling Brain* (1st ed.). Orlando: Harcourt.

Davidson, R. J. (2000). Cognitive Neuroscience Needs Affective Neuroscience (and Vice Versa). *Brain and Cognition, 42*(1), 89–92.

Davis, J. (2009). Design Methods for Ethical Persuasive Computing. In *Proceedings of the 4th International Conference on Persuasive Technology* (pp. 6–13). New York: ACM.

Dawson, M. (2003). *The Consumer Trap: Big Business Marketing in American Life.* Urbana: University of Illinois Press.

de Oliveira, R., & Carrascal, J. P. (2014). Towards Effective Ethical Behavior Design. In *Proceedings of CHI '14* (pp. 2149–2154). New York: ACM.

DiSalvo, C., Sengers, P., & Brynjarsdóttir, H. (2010). Mapping the Landscape of Sustainable HCI. In *Proceedings of CHI '10* (pp. 1975–1984). New York: ACM.

Dourish, P. (2010). HCI and Environmental Sustainability: The Politics of Design and the Design of Politics. In O. W. Bartelsen & P. Krogh (Eds.), *Proceedings of DIS 2010* (pp. 1–10). New York: ACM.

Egan, P. J., & Mullin, M. (2017). Climate Change: US Public Opinion. *Annual Review of Political Science, 20,* 209–227.

Ewen, S. (2001). *Captains of Consciousness: Advertising and the Social Roots of the Consumer Culture* (25th anniversary ed.). New York: Basic Books.

Fogg, B. J. (2003). *Persuasive Technology: Using Computers to Change What We Think and Do.* Amsterdam/Boston: Morgan Kaufmann Publishers.

Foster, D., Lawson, S., Blythe, M., & Cairns, P. (2010). Wattsup?: Motivating Reductions in Domestic Energy Consumption Using Social Networks. In *Proceedings of NordiCHI 2010* (pp. 178–187). New York: ACM.

Friedman, B. (1996). Value-Sensitive Design. *Interactions, 3*(6), 17–23.

Friedman, B., & Nissenbaum, H. (1996). Bias in Computer Systems. *ACM Transactions on Information Systems, 14*(3), 330–347.

Froehlich, J., Findlater, L., & Landay, J. (2010). The Design of Eco-feedback Technology. In *Proceedings of CHI 2010* (pp. 1999–2008). New York: ACM.

Gabrielli, S., & Maimone, R. (2014). Designing a Context-Aware Mobile Application for Eco-driving. In *Proceedings of ICCASA 14* (pp. 102–104). Brussles: ICST.

Gamberini, L., Spagnolli, A., Corradi, N., Jacucci, G., Tusa, G., Mikkola, T., et al. (2012). Tailoring Feedback to Users' Actions in a Persuasive Game for Household Electricity Conservation. In M. Bang & E. L. Ragnemalm (Eds.), *Persuasive Technology. Design for Health and Safety. PERSUASIVE 2012* (pp. 100–111). Berlin/Heidelberg: Springer.

Ganglbauer, E., Reitberger, W., & Fitzpatrick, G. (2013). An Activist Lens for Sustainability: From Changing Individuals to Changing the Environment. In S. Berkovsky & J. Freyne (Eds.), *PERSUASIVE 2013* (pp. 63–68). Berlin/Heidelberg: Springer.

Gauntlett, D. (2005). *Moving Experiences: Media Effects and Beyond* (2nd ed.). Eastleigh/Bloomington: John Libbey Pub.

Geelen, D., Keyson, D., Boess, S., & Brezet, H. (2012). Exploring the Use of a Game to Stimulate Energy Saving in Households. *Journal of Design Research, 10*(1–2), 102–120.

Gregory, J., & Miller, S. (1998). *Science in Public: Communication, Culture, and Credibility*. New York: Perseus.

Guerin, D. A., Yust, B. L., & Coopet, J. G. (2000). Occupant Predictors of Household Energy Behavior and Consumption Change as Found in Energy Studies Since 1975. *Family and Consumer Sciences Research Journal, 29*(1), 48–80.

Hamari, J., Koivisto, J., & Pakkanen, T. (2014). Do Persuasive Technologies Persuade? – A Review of Empirical Studies. In A. Spagnolli, L. Chittaro, & L. Gamberini (Eds.), *Persuasive Technology. PERSUASIVE 2014* (pp. 137–142). Cham: Springer.

Hassenzahl, M., & Laschke, M. (2015). Pleasurable Troublemakers. In S. P. Walz & S. Deterding (Eds.), *The Gameful World: Approaches, Issues, Applications* (pp. 167–195). London/Cambridge, MA: MIT Press.

Hazas, M., Bernheim Brush, A. J., & Scott, J. (2012). Sustainability Does Not Begin with the Individual. *Interactions, 19*(5), 14–17.

Hiraoka, T., Terakado, Y., Matsumoto, S., & Yamabe, S. (2009). Quantitative Evaluation of Eco-driving on Fuel Consumption Based on Driving Simulator Experiments. In *Proceedings of the 16th World Congress on Intelligent Transport Systems* (pp. 21–25). Washington, DC: ITS.

Hornsey, M. J., Harris, E. A., Bain, P. G., & Fielding, K. S. (2016). Meta-Analyses of the Determinants and Outcomes of Belief in Climate Change. *Nature Climate Change, 6*, 622.

Hulme, M. (2009). *Why We Disagree About Climate Change: Understanding Controversy, Inaction and Opportunity*. Cambridge/New York: Cambridge University Press.

Kahan, D. M., Jenkins-Smith, H., & Braman, D. (2011). Cultural Cognition of Scientific Consensus. *Journal of Risk Research, 14*(2), 147–174.

Kahneman, D. (2011). *Thinking, Fast and Slow*. New York: Farrar, Straus and Giroux.

Karppinen, P., & Oinas-Kukkonen, H. (2013). Three Approaches to Ethical Considerations in the Design of Behavior Change Support Systems. In S. Berkovsky & J. Freyne (Eds.), *PERSUASIVE 2013* (pp. 87–98). Berlin/Heidelberg: Springer.

Kim, P. (1992). Does Advertising Work: A Review of the Evidence. *Journal of Consumer Marketing, 9*(4), 5–21.

Kim, T., Hong, H., & Magerko, B. (2010). Design Requirements for Ambient Display That Supports Sustainable Lifestyle. In *Proceedings of DIS 2010* (pp. 103–112). New York: ACM.

Klein, N. (2014). *This Changes Everything: Capitalism vs. the Climate*. Toronto: Alfred A. Knopf.

Knowles, B., Blair, L., Coulton, P., & Lochrie, M. (2014). Rethinking Plan A for Sustainable HCI. In *Proceedings of CHI 2014* (pp. 3593–3596). New York: ACM.

Koger, S. M., & Winter, D. D. N. (2010). *The Psychology of Environmental Problems: Psychology for Sustainability* (3rd ed.). New York/London: Psychology Press.

Kollmuss, A., & Agyeman, J. (2002). Mind the Gap: Why Do People Act Environmentally and What Are the Barriers to Pro-environmental Behavior? *Environmental Education Research, 8*(3), 239–260.

Krimsky, S., & Plough, A. L. (1988). *Environmental Hazards: Communicating Risks as a Social Process.* Dover: Auburn House.

Kuijer, L., & Bakker, C. (2015). Of Chalk and Cheese: Behaviour Change and Practice Theory in Sustainable Design. *International Journal of Sustainable Engineering, 8*(3), 219–230.

Lakoff, G., & Johnson, M. (1999). *Philosophy in the Flesh: The Embodied Mind and Its Challenge to Western Thought.* New York: Basic Books.

Lavender, T. (2010, July 18). Digital Media Students Want to Raise Your Carbon Consciousness. *Vancouver Observer.* Retrieved from http://www.vancouverobserver.com/blogs/megabytes/2010/07/18/digital-media-students-want-raise-your-carbon-consciousness

Lee, H., Lee, W., & Lim, Y.-K. (2010). The Effect of Eco-driving System Towards Sustainable Driving Behavior. In *Proceedings of CHI 2010* (pp. 4255–4260). New York: ACM.

Lilley, D., & Wilson, G. T. (2013). Integrating Ethics into Design for Sustainable Behaviour. *Journal of Design Research, 11*(3), 278–299.

Lilley, D., Lofthouse, V., & Bhamra, T. (2005). Towards Instinctive Sustainable Product Use. Paper Presented at the 2nd International Conference: Sustainability Creating the Culture, Aberdeen.

Lockton, D. (2013). *Design with Intent: A Design Pattern Toolkit for Environmental & Social Behaviour Change.* PhD dissertation submitted at Brunel University, London.

Losh, E. M. (2009). *Virtualpolitik: An Electronic History of Government Media-Making in a Time of War, Scandal, Disaster, Miscommunication, and Mistakes.* Cambridge, MA: MIT Press.

Lövbrand, E., Beck, S., Chilvers, J., Forsyth, T., Hedrén, J., Hulme, M., et al. (2015). Who Speaks for the Future of Earth? How Critical Social Science Can Extend the Conversation on the Anthropocene. *Global Environmental Change, 32*, 211–218.

Lupia, A., McCubbins, M. D., & Popkin, S. L. (2000). Beyond Rationality: Reason and the Study of Politics. In A. Lupia, M. D. McCubbins, & S. L. Popkin (Eds.), *Elements of Reason: Cognition, Choice, and the Bounds of Rationality* (pp. 1–20). Cambridge/New York: Cambridge University Press.

Mankoff, J. C., Fussell, S. R., Dillahunt, T., Glaves, R., Grevet, C., Johnson, M., et al. (2010). StepGreen.org: Increasing Energy Saving Behaviors Via Social Networks. In *Proceedings of the Fourth International AAAI Conference on Weblogs and Social Media* (pp. 106–113). Palo Alto: AAAI.

Mazé, R. (2010). *Static! Designing for Energy Awareness*. Stockholm: Arvinius Förlag.

McKenzie-Mohr, D. (2000). Fostering Sustainable Behavior Through Community-Based Social Marketing. *American Psychologist, 55*(5), 531–537.

McRae, L., Freeman, R., & Marconi, V. (2016). The Living Planet Index. In N. Oerlemans (Ed.), *Living Planet Report 2016: Risk and Resilience in a New Era*. Gland: WWF International.

Niedderer, K., Ludden, G., Clune, S. J., Lockton, D., Mackrill, J., Morris, A., et al. (2016). Design for Behaviour Change as a Driver for Sustainable Innovation: Challenges and Opportunities for Implementation in the Private and Public Sectors. *International Journal of Design, 10*(2), 67–85.

Nodder, C. (2013). *Evil by Design: Interaction Design to Lead Us into Temptation*. Indianapolis: Wiley.

Nyhan, B., & Reifler, J. (2010). When Corrections Fail: The Persistence of Political Misperceptions. *Political Behavior, 32*(2), 303–330.

Oinas-Kukkonen, H. (2013). A Foundation for the Study of Behavior Change Support Systems. *Personal and Ubiquitous Computing, 17*(6), 1223–1235.

Oinas-Kukkonen, H., & Harjumaa, M. (2009). Persuasive Systems Design: Key Issues, Process Model, and System Features. *Communications of the Association for Information Systems, 24*(article 28), 485–500.

Packard, V. (1957). *The Hidden Persuaders*. New York: D. McKay Co.

Peffer, T., Pritoni, M., Meier, A., Aragon, C., & Perry, D. (2011). How People Use Thermostats in Homes: A Review. *Building and Environment, 46*(12), 2529–2541.

Pettersen, I. N., & Boks, C. (2008). The Ethics in Balancing Control and Freedom When Engineering Solutions for Sustainable Behaviour. *International Journal of Sustainable Engineering, 1*(4), 287–297.

Pierce, J., Odom, W., & Blevis, E. (2008). Energy Aware Dwelling: A Critical Survey of Interaction Design for Eco-visualizations. Paper Presented at OZCHI 2008, Cairns.

Prost, S., Schrammel, J., & Tscheligi, M. (2014). 'Sometimes It's the Weather's Fault': Sustainable HCI & Political Activism, CHI '14 Extended Abstracts on Human Factors in Computing Systems (pp. 2005–2010). New York: ACM.

Rozendaal, M. (2016). Objects with Intent: A New Paradigm for Interaction Design. *Interactions, 23*(3), 62–65.

Ruijten, P. A. M., Midden, C. J. H., & Ham, J. (2011). Unconscious Persuasion Needs Goal-Striving: The Effect of Goal Activation on the Persuasive Power of Subliminal Feedback. In *Proceedings of Persuasive 2011 (article number 4)*. New York: ACM.

Rutherford, A. (2003). B. F. Skinner's Technology of Behavior in American Life: From Consumer Culture to Counterculture. *Journal of History of the Behavioral Sciences, 39*(1), 1–23.

Samuel, L. R. (2010). *Freud on Madison Avenue: Motivation Research and Subliminal Advertising in America*. Philadelphia: University of Pennsylvania Press.

Shove, E. (2010a). Beyond the ABC: Climate Change Policy and Theories of Social Change. *Environment and Planning A, 42*, 1273–1285.

Shove, E. (2010b). Social Theory and Climate Change: Questions Often, Sometimes and Not Yet Asked. *Theory, Culture & Society, 27*(2–3), 277–288.

Shove, E. (2011). On the Difference Between Chalk and Cheese – A Response to Whitmarsh et al.'s Comments on 'Beyond the ABC: Climate Change Policy and Theories of Social Change'. *Environment and Planning A, 43*, 262–264.

Shove, E., & Spurling, N. (Eds.). (2013). *Sustainable Practices: Social Theory and Climate Change*. London/New York: Routledge.

Skinner, B. F. (1971). *Beyond Freedom and Dignity*. Indianapolis/Cambridge: Hackett.

Slovic, P. (1987). Perception of Risk. *Science, 236*(4799), 280–285.

Smids, J. (2012). The Voluntariness of Persuasive Technology. In M. Bang & E. L. Ragnemalm (Eds.), *PERSUASIVE 2012* (pp. 123–132). Heidelberg: Springer.

Snow, S., Buys, L., Roe, P., & Brereton, M. (2013). Curiosity to Cupboard: Self Reported Disengagement with Energy Use Feedback Over Time. In *Proceedings of OzCHI '13* (pp. 245–254). New York: ACM.

Stern, P. C. (2000). Toward a Coherent Theory of Environmentally Significant Behavior. *Journal of Social Issues, 56*(3), 407–424.

Strengers, Y. (2011). Designing Eco-feedback Systems for Everyday Life. In *Proceedings of CHI 2011* (pp. 2135–2144). New York: ACM.

Strengers, Y., & Maller, C. (Eds.). (2015). *Social Practices, Intervention and Sustainability: Beyond Behaviour Change*. London/New York: Routledge.

Strike, K. A. (1975). Beyond Freedom and Dignity. *Studies in Philosophy and Education, 9*(1–2), 112–137.

Tavris, C., & Aronson, E. (2007). *Mistakes Were Made (But Not by Me): Why We Justify Foolish Beliefs, Bad Decisions, and Hurtful Acts*. Orlando: Harcourt.

Taylor, C. (2004). *Modern Social Imaginaries*. Durham: Duke University Press.

Thaler, R. H. (2000). From Homo Economicus to Homo Sapiens. *Journal of Economic Perspectives, 14*(1), 133–141.

Thaler, R. H., & Sunstein, C. R. (2008). *Nudge: Improving Decisions About Health, Wealth, and Happiness*. New Haven: Yale University Press.

Thieme, A., Comber, R., Miebach, J., Weeden, J., Krämer, N., Lawson, S., & Olivier, P. (2012). "We've Bin Watching You": Designing for Reflection and Social Persuasion to Promote Sustainable Lifestyles. In *Proceedings of CHI 2012* (pp. 2337–2346). New York: ACM.

Timmer, J., Kool, L., & van Est, R. (2015). Ethical Challenges in Emerging Applications of Persuasive Technology. In T. MacTavish & S. Basapur (Eds.), *Persuasive Technology. PERSUASIVE 2015* (pp. 196–201). Cham: Springer.

Tromp, N., Hekkert, P., & Verbeek, P.-P. (2011). Design for Socially Responsible Behavior: A Classification of Influence Based on Intended User Experience. *Design Issues, 27*(3), 3–19.

Tversky, A., & Kahneman, D. (1982). Judgment Under Uncertainty: Heuristics and Biases. In D. Kahneman, P. Slovic, & A. Tversky (Eds.), *Judgment Under Uncertainty: Heuristics and Biases* (pp. 3–20). Cambridge/New York: Cambridge University Press.

Vaughn, R. (1980). How Advertising Works: A Planning Model. *Journal of Advertising Research, 20*(5), 27–33.

Verbeek, P.-P. (2006). Persuasive Technology and Moral Responsibility: Toward an Ethical Framework for Persuasive Technologies. Paper Presented at Persuasive 06, Eindhoven, the Netherlands.

Wallace-Wells, D. (2017, July 9). The Uninhabitable Earth. *New York Magazine.* Retrieved from http://nymag.com/daily/intelligencer/2017/07/climate-change-earth-too-hot-for-humans.html

Whitmarsh, L., O'Neill, S., & Lorenzoni, I. (2011). Climate Change or Social Change? Debate Within, Amongst, and Beyond Disciplines. *Environment and Planning A, 43*, 258–261.

Wilson, G. T., Lilley, D., & Bhamra, T. A. (2013). Design Feedback Interventions for Household Energy Consumption Reduction. Paper Presented at the ERSCP-EMSU 2013 Conference, Istanbul.

Wilson, G. T., Bhamra, T., & Lilley, D. (2015). The Considerations and Limitations of Feedback as a Strategy for Behaviour Change. *International Journal of Sustainable Engineering, 8*(3), 186–195.

Wynne, B. (1992). Misunderstood Misunderstanding: Social Identities and Public Uptake of Science. *Public Understanding of Science, 1*(3), 281–304.

Yang, R., Newman, M. W., & Forlizzi, J. (2014). Making Sustainability Sustainable: Challenges in the Design of Eco-interaction Technologies. In *Proceedings of CHI 2014* (pp. 823–832). New York: ACM.

Zachrisson, J., & Boks, C. (2012). Exploring Behavioural Psychology to Support Design for Sustainable Behaviour Research. *Journal of Design Research, 10*(1–2), 50–66.

Zapico, J. L., Turpeinen, M., & Brandt, N. (2009). Climate Persuasive Devices: Changing Behaviour Towards Low-Carbon Lifestyles. In S. Chatterjee & P. Dev (Eds.), *Proceedings of Persuasive '09* (Article 14). New York: ACM Press.

Zappen, J. P. (2005). Digital Rhetoric: Toward an Integrated Theory. *Technical Communication Quarterly, 14*(3), 319–325.

CHAPTER 3

Reason

THE WORLD GAME

R. Buckminster Fuller, the visionary polymath, had a dream. In 1964 he suggested that the United States Information Agency would support his creation of a computerized world simulation—an interactive global dashboard that would be the centerpiece of the American pavilion to Expo 67 in Montreal. In Fuller's vision, visitors would enter the pavilion through 36 ramps and escalators to encounter a model world globe, 100 feet in diameter, suspended high in the air. As they gather around the quarter-mile-long balcony that runs the perimeter of the building (a geodesic dome, naturally), the massive globe would open up and flatten out to become a giant Dymaxion map.[1] The map would then be populated with a multitude of mini light bulbs that indicate "various, accurately positioned, proportional data regarding world conditions, events, and resources" (Fuller 1969, p. 112), allowing visitors, as individuals or in teams, to play a "great world logistics game."

Foreshadowing the Project on the Predicament of Mankind (whose results would later be published as *The Limits to Growth* report (Meadows

---

[1] A Dymaxion Map is a "cartographic projection" that allows a view of "the whole of the Earth's surface simultaneously without any visible distortion of the relative size and shape of the land and sea masses having occurred during the transformation from sphere to the flattened-out condition which we call a map" (Fuller 1969, p. 111).

© The Author(s) 2018
R. Bendor, *Interactive Media for Sustainability*,
Palgrave Studies in Media and Environmental Communication,
https://doi.org/10.1007/978-3-319-70383-1_3

et al. 1972)),[2] the *World Game* would be run by computers loaded with "all the known inventory and whereabouts of the various metaphysical and physical resources of the Earth," and "all the inventory of human trends, known needs and fundamental behavior characteristics" (Fuller 1969, p. 112)—an inventory Fuller had started to assemble nearly 40 years earlier.[3] Additional data would be collected from satellite sensors, weather reports, crop patterns, census data, and other sources, resulting in a sophisticated computer-based simulation capable of dynamically tracing and making visible the entities, interrelations, and processes that underlie the planet's complex metabolic systems. As Fuller recounts,

> I proposed that individuals and teams would undertake to play the 'World Game' with those resources, behaviors, trends, vital needs, developmental desirables, and regenerative inspirations. The players as individuals or teams would each develop their own theory of how to make the total world work successfully for all of humanity. Each individual or team would play his theory through to the end of his predeclared program. It could be played with or without competitors.

> The objective of the game would be to explore for ways to make it possible for anybody and everybody in the human family to enjoy the total earth without any human interfering with any other human and without any human gaining advantage at the expense of another. (Fuller 1969, p. 112)

Playing the game, Fuller adds, will provide humanity with the means to override political shortsightedness and self-interest with the assistance of computerized reason, converting "the trend of humanity toward extinction into a trend toward universal physical and metaphysical success" (ibid., p. 118). To win the game, "everybody must be made physically successful. *Everybody* must win" (ibid., p. 114; emphasis added). In this sense, the game embodies Fuller's belief that human needs and aspirations are inseparable from natural phenomena and processes, comprising together a whole that is far greater than the sum of its parts. Painting a picture of sustainability as more than a zero-sum game—"A world-around game of musical chairs" as Fuller (ibid.) puts it—Fuller's position

---

[2] The report cites Fuller as an example of fanciful technological optimism (Meadows et al. 1972, p. 130). Fuller, in response, would later criticize the report as "Malthusian" (see Fuller 1999, p. 38).

[3] See introduction to Fuller (1969, p. vi).

anticipates what would later be called "regenerative sustainability" (Robinson and Cole 2015).

Disappointingly, Fuller's *World Game*, "the capstone of a remarkable life," as T.B. Turner writes in his foreword to Fuller (1969, p. iv), was never built.[4] But it does make visible the contours of what I call here *synoptic interactions*. These are mediated interactions that aim to convey sustainability from a systemic, holistic perspective, rendering the complex interrelations that underlie sustainability tangible and thus malleable.[5] As I discuss below, interactive media that deploy synoptic interactions promote the development of appropriate mental models of sustainability by providing users with the means to interact with spatiotemporal computational models that simulate socioenvironmental systems. Much in the spirit of Fuller's *World Game*, this implies an unyielding belief that humans are capable of grasping, navigating, and acting on complex systems, but also that computation is capable of augmenting human cognition in ways that ameliorate innate biases, shortsightedness, informational gaps, and, no less important, ideological predispositions. Fuller certainly believed in the transformative power of computational simulations. He wrote, "I am quite confident that as the world game is played progressively it will disclose a myriad of politically untried, unprecedented yet amazingly effective ways of solving hitherto unsurmountable world-around problems" (Fuller 1969, p. 116). Seen in this light, synoptic interactions articulate a particular configuration of epistemological commitments and political intentionality. How they do so is the subject of this chapter.

From Complexity to Wickedness

The object of sustainability is the world. And as the natural and social sciences tirelessly remind us, the world is incredibly complex and riddled with uncertainty. The claim of ecology, for instance, is that nature involves

---

[4]A pilot version of the game was played by students at the New York Studio School of Painting and Sculpture in 1969 (see timeline in Fuller (1999, p. 37)). According to Sisson (2015, Dec. 14), it was later staged in schools across the United States. A modernized version of the game is offered by o.s.Earth as an educational tool (see http://www.osearth.com/the-game (last accessed Mar. 18, 2018)).

[5]I first proposed what I call here synoptic interactions as "analytic interactions" in Bendor (2012). However, that initial term did not capture the way such interactions aim not to break complex systems into their components (the *modus operandi* of analysis) but to grasp them holistically—to provide a "synoptic grasp of the ensemble" as Scott (1998, p. 59) puts it.

such a degree of interconnectedness that no natural entity can be fully understood in isolation from other members of its ecosystem (Odum 1983). But while early ecological models viewed natural systems as essentially tending to orderly equilibrium, scientists now espouse a view of nature that "is far from equilibrium, nonlinear and full of irreversible processes" (Prigogine, cited in Connolly 2006, p. 71), "evolving, contingent, revolutionary, conflicted, catastrophic at times, always in a state of flux" (Worster 1994, p. 421).[6] The natural world is understood as the integrated outcome of a multitude of autonomous or semi-autonomous, nested or interconnected, linear, nonlinear, or chaotic systems: "each system – when examined in the timescale appropriate to it – oscillates between periods of relative arrest and heightened imbalance and change, followed in turn by new stabilizations, some of which may assume a composition never fully manifest before" (Connolly 2006, p. 70). In a similar vein, Bill Rees (2012, p. 251) writes,

> The ecosphere ... is a highly-ordered self-organizing system of mind-boggling complexity, multi-layered structure and steep gradients as represented by millions of distinct species, differentiated matter, and accumulated biomass. Over geological time, its biodiversity, systemic complexity, and energy/material flows have been increasing – i.e., the ecosphere has been moving ever *further* from equilibrium. (emphasis in origin)

To identify equilibria (or the lack of), we first need to recognize patterns, yet while human cognition seeks balance and simple patterns in the environment—patterns that evidence causality and allow the modelling and control (or management) of environmental processes—the environment itself exhibits a confounding tendency for nonlinear complexity and emergent processes with unpredictable outcomes.[7] The patterns keep changing. Nature, Donald Worster points out, "is *fundamentally* erratic, discontinuous, and unpredictable.... As a result, the unexpected keeps hitting us in the face" (cited in Merchant 2008, p. 381; emphasis in origin). For evidence of how nature's erratic complexity repeatedly "hits us in the face," we need to look no further than to climate change and the extreme weather events it catalyzes. But when it comes to sustainability, the kind of nonlinear complexity exhibited by nature is infinitely compounded

---

[6] See also Scott (2007), and essays by Capra, Worster, Bohm, and Prigogine in Merchant (2008).

[7] Fritjof Capra refers to this gap as "a crisis of perception" (in Merchant 2008, p. 365).

by strongly interrelating human social, cultural, technological, and economic phenomena (Dyball and Newell 2015; Espinosa and Walker 2011; Levin 1999; Westley et al. 2002). Complexity becomes, as Peter Taylor (2005) writes, "unruly."

In a world characterized by complexity and uncertainty, and where social and natural systems interact with unanticipated results, sustainability appears as a prototypical "wicked" problem. Wicked problems, explain Rittel and Webber (1973) in a seminal article, are ill defined, hard to analytically isolate by separating them from other problems, and stubbornly resistant to permanent solutions. As such, they reflect the limitations of social analysis and decision-making in a social reality girded by a pluralism of values and publics, the consequence of which are that "The professionalized cognitive and occupational styles that were refined in the first half of this [twentieth] century, based in Newtonian mechanistic physics, are not readily adapted to contemporary conceptions of interacting open systems and to contemporary concerns with equity" (ibid., p. 156). Our previous models of complexity became obsolete in the wake of new observations about the degree to which the world is complex. Since a wicked problem's definition is co-emergent with its solution—"in order to *describe* a wicked-problem in sufficient detail, one has to develop an exhaustive inventory of all conceivable *solutions* ahead of time" (ibid., p. 161; emphasis in origin)—we can never really know whether it was actually solved. In fact, Rittel and Webber state, work on wicked problems is never concluded "for reasons inherent in the 'logic' of the problem," but instead "for considerations that are external to the problem" like running out of time, money, or patience (ibid., p. 162). In other words, wicked problems are never solved but *re-solved* again and again. And because wicked problems inherently include normative elements, we lack objective criteria for evaluating solutions. As Taylor (2005) writes, when encountering unruly complexity, "no privileged standpoint exists" (p. 157). Navigating wicked problems, it follows, relies on relative (or comparative) judgment; solutions cannot be seen as "true" or "false" but as "better" or "worse" (Rittel and Webber 1973, p. 163). Wicked problems are co-equivalent with an emergent problem space that is in constant flux, where facts and values intermingle, and where stability can only be found in the anticipation of its opposite.

How can we solve wicked problems such as sustainability? Or, more accurately, how can we manage them in a way that prevents the Anthropocene from becoming humanity's epitaph? As mentioned above, while some sustainability issues may lend themselves to fairly straightforward technical solutions, the addition of human interests and politics into the mix creates a challenge of a whole different class, effectively turning complexity into wickedness. A possible strategy for dealing with wicked problems is to understand them as complex systems—to *"render a complex situation system-like"* (Taylor 2005, p. 157; emphasis in origin). Such a strategy often makes use of resources provided by systems thinking: an epistemology and a methodology (as is often the case) that "gives us the freedom to identify root causes of problems and see new opportunities" (Meadows 2008, p. 2). Systems thinking, much like its forerunners Tektology, ecology, cybernetics, and general systems theory, is both holistic and relational, that is, it sees the relations that bind entities as ontologically preceding those very entities. In Fritjof Capra's (1985, p. 476) words, "The whole is primary," which is to say that systems strongly influence the fundamental conditions within which their elements exist and function. Donella Meadows, famed lead author of the *Limits to Growth* report, one of the most influential examples of systems thinking in action, points out that "The behavior of a system cannot be known just by knowing the elements of which the system is made" (2008, p. 7). In this sense,

> there is an integrity or wholeness about a system and an active set of mechanisms to maintain that integrity. Systems can change, adapt, respond to events, seek goals, mend injuries, and attend to their own survival in lifelike ways, although they may contain or consist of nonliving things. Systems can be self-organizing, and often are self-repairing over at least some range of disruptions. They are resilient, and many of them are evolutionary. Out of one system other completely new, never-before-imagined systems can arise. (ibid., p. 12)

The founder of system dynamics, Jay Forrester (1998), provides a similar evaluation:

> We do not live in a unidirectional world in which a problem leads to an action that leads to a solution. Instead, we live in an on-going circular environment. Each action is based on current conditions, such actions affect future conditions, and changed conditions become the basis for later action. There is no beginning or end to the process. (pp. 2–3)

Accordingly, system thinking's methodology involves a *continuous*, at times Sisyphean, assessment of parts and wholes—a sustained focus on entities, their interrelations, and the function or purpose of the system *as a whole*.

Aside from entities and relations, every system has "stocks" (material and other assets), which fluctuate in response to "flows" (inputs and outputs).[8] The relation between stocks and flows is regulated by feedback processes or "loops," which can balance the system or throw it into "runaway" collapse. The crucial factor, however, is how the system adapts to change: it can exhibit "resilience" (the degree to which it is capable of bouncing back from disruptive events), "self-organization" (a capacity to diversify, complexify and evolve), and "hierarchy" (the way in which it interacts with other, encompassing or nested systems). Of these, self-organization is perhaps the most important characteristic for it illustrates complex systems as emergent phenomena, that is, phenomena that exhibit "emergence": "The movement from low-level rules to higher-level sophistication" (Johnson 2001, p. 18). Simply stated, emergence references how complex systems adapt, change, or morph in ways that "are neither predictable from, deducible from, nor reducible to the parts alone" (Goldstein 1999, p. 50).

Systems thinking offers those seeking to engage with wicked problems a useful conceptual vocabulary with which to contend with, appreciate, and even relish the messiness of the world. Take, for example, Meadows's (2008, ch.7) summary of systems thinking imperatives: settle not for simplified causalities ("Celebrate complexity"), be reflexive ("Expose your mental models to the light of day"), avoid fixating on quantifiable phenomena ("Pay attention to what is important, not just what is quantifiable"), and consider multiple temporalities ("Expand time horizons"). But with advice such as "Stay humble – stay a learner," "listen to the wisdom of the system," "Expand the boundary of caring," and "Don't erode the goal of goodness," we can see how systems thinking also manifests a philosophy—an almost Zen-like way of being-in-the-world[9]: "We can't impose our will on a system. We can listen to what the system tells us, and discover how its properties and our values can work together to bring forth something much better than could ever be produced by our will alone" (Meadows 2008, pp. 169–170). It is apparent, then, that what makes systems thinking innovative when compared to its forerunners is precisely that which makes it compatible for engaging wicked problems: it treats equilibria and homeostasis as merely temporary states in what is otherwise an emergent world in perpetual flux. No

[8] I am relying here on Meadows's excellent *Thinking in Systems: A Primer* (2008). For a more comprehensive account of systems thinking, see Midgley (2003).

[9] See also Lilienfeld (1975).

stable foundations, just "a dynamic web of interrelated events" (Capra 1985, p. 477) that is infinitely complicated by the active presence of social phenomena. This is perhaps most pronounced in the way systems thinking integrates into its very core "surprising events"—those instances where our mental models of reality are incongruent with what we actually perceive. Nonlinear processes, nonexistent boundaries, disruptive temporalities, and the inescapable effects of bounded rationality are all fundamental to the way complex systems are addressed; not hurdles to overcome on the path toward systems mastery but elements of a messy world that is, nonetheless, our reality. Again, in Meadows's (2008) words: "Let's face it, the universe is messy. It is nonlinear, turbulent, and dynamic.... It self-organizes and evolves. It creates diversity *and* uniformity. That's what makes the world interesting, that's what makes it beautiful, and that's what makes it work" (p. 181; emphasis in origin). Encoding, communicating, and rendering actionable this "beautiful messiness" is the main goal of synoptic interactions.

Playing Sustainability with the Sims

In February 2014, six groups of urban thinkers were brought together for a friendly exercise in city-building with the urban simulation game *SimCity*. The game, a digital sandbox powered by simple artificial intelligence—"a hard core of modeling simulation in a soft shell of gaming" as Mayer (2008, p. 848) writes in a different context—spawned a successful franchise currently owned by gaming giant Electronic Arts (EA). It is the brainchild of Will Wright, who started to work on it in 1985.[10] The first version was released in 1989, and has since spawned numerous variations and spinoffs into what some consider "the best-selling PC game franchise ever."[11]

Based on a sophisticated simulation engine, the game allows players to create single cities or entire regions populated with artificial intelligence agents: Sims. City-building includes a large variety of spatial or geographic features.[12] Players can apply zoning regulations and build roads, homes,

---

[10] https://www.ea.com/en-ca/games/simcity (last accessed Mar. 18, 2018).

[11] '2016 World Video Game Hall of Fame Inductees Announced: *Grand Theft Auto III, The Legend of Zelda, The Oregon Trail, The Sims, Sonic the Hedgehog, and Space Invaders*': http://www.museumofplay.org/press/releases/2016/05/2688-2016-world-video-game-hall-fame-inductees-announced (last accessed Mar. 18, 2018).

[12] The description that follows is based on *SimCity* 2013, widely recognized as significantly more "sustainability friendly" than previous versions. *Slate* magazine's reviewer called it "the best urban-planning simulation ever created" (Manjoo 2013, Mar. 4).

businesses, industrial facilities, and public utilities including water, sewage, and transit. They can also power their city from a variety of renewable and fossil fuel-based energy sources. By making decisions about resources, infrastructure, and services, players manipulate a variety of social and economic activities pertaining to employment, education, and tourism and respond to both economic imperatives such as financial balance and social imperatives such as level of policy approval. The consequences are quite realistic. The Sims may prosper and the city grow, or pollution, gridlocked traffic, and a declining economy may bring about health problems, rising crime, and dissatisfaction. These, and other indicators such as overall costs, population size, power production and demand, land value, urban density, radiation and air pollution, and even the Sims's level of happiness, are visible to players on a dashboard that updates in real time. Furthermore, by speeding up or pausing the game, players can get a better sense of the impact of their choices, and the game's simulation of natural disasters such as earthquakes and tornados—and even Godzilla-like monsters that invade the city!—provides a dramatic test of the city's resilience. The game, in other words, simulates a complex urban system along with all of its typical components: entities, interrelations, stocks, flows, and feedback loops.[13] With the addition of autonomously acting agents for every in-game entity (cars, people, etc.), the game even exhibits emergence (Lorince 2013, Mar. 6). It is no surprise, then, that the game's pedagogical appeal and utility have been debated in the context of urban geography and planning (Adams 1998; Bereitschaft 2016; Devisch 2008; Gaber 2007; Kim and Shin 2016; Nilsson and Jakobsson 2011; Terzano and Morckel 2016), and even political science (Woessner 2015). A version designed especially for educational purposes (grades 6–8) is currently offered by GlassLab: "SimCityEDU: Pollution Challenge! not only teaches students about the factors affecting the environment in a modern city, but the game also provides formative assessment information about students' ability to problem solve and explain the relationships in complex systems".[14]

Returning to the city-building competition, the event was organized by business media firm Fast Company and included teams from MIT's

---

[13] Seeing *SimCity* as a manifestation of systems thinking is not serendipitous. Will Wright, the game's inventor, acknowledged his debt to Jay Forrester's work on system dynamics and urban planning (Mayer 2008, p. 837; Wells 2016, p. 529).

[14] 'SimCityEDU: Pollution Challenge!': http://www.glasslabgames.org/games/SC (last accessed Mar. 18, 2018).

Department of Urban Study and Planning, Columbia University's future cities laboratory Studio-X, architecture firms Kohn Pedersen Fox (KPF) and Studio Gang, the nonprofit city planning firm OpenPlans, and Fast Company itself (McDermott 2014, Feb. 9).[15] There were no strict rules (beyond the game's), as teams "were merely encouraged to create what they felt was the most well-structured urban environment." Accordingly, each team operationalized its own driving principles (a system's "purpose" in the language of systems thinking). These ranged from a "walkable, eco-friendly city with minimal car traffic" (OpenPlan's *Openopolis*) and "a dense commercial and residential hub" (KPF's *XimCity*), to a city based on "a knowledge economy" (Studio-X's *Champignon*). Unintentionally yet consistently, all simulated cities practiced economic autonomy, that is, there were no regional collaborations beyond market-like transactions in goods and services.

The rationale behind the competition was to see what urban specialists could do with *SimCity*'s powerful simulation engine. As the organizers remark, "The thought was that coupling the players' collective genius with *SimCity*'s planning dashboard would result in a vision of a potential urban future, a blueprint for the future of cities." Alas, "That would not turn out to be the case." This is plainly evident from the descriptions of the final products—"Think Manhattan, but with casinos and wind farms"; "One side of the city is dedicated to the oil industry. The other end has a bustling downtown area. Also, there's a pretty massive convention center"; "oceanside wonder that's filled with casinos, cops and beach houses"; "A radial city that aspires to be a cosmopolitan metropolis but more closely resembles Scottsdale, Ariz."; and "An oil and ore town built upon a topographical peninsula. The driving philosophy is unabashed capitalism." Not exactly shining models of urban sustainability. But what makes the competition so interesting is not the unfortunate perversion of the teams' original sustainable goals, but the reasons why teams ended up making unsustainable choices: "Reflecting later, the urbanists were most concerned about the game's value system. By encouraging players to make short-sighted decisions, *SimCity* trivializes the importance of certain public policies (e.g. investing in education, renewable energy, and pedestrianism)." The game, it seems, was designed to promote unsustainable choices, a fact confirmed by its lead designer, Stone Librande: "It's designed to make players make unsustainable decisions. We want people to understand

---

[15] Additional references to the competition are taken from the same article.

why it happens in the real world ... If the game pulls you into this path that you know is bad and you know is wrong, you start to understand why we do things like mountaintop removal to get coal" (cited in McDermott 2014, Feb. 9).

The game's proclivity for unsustainability was reflected in the in-game availability and accessibility of fossil fuels, in the unavailability of certain design and policy options, but also in the dynamics of its feedback loops. Librande, again, confirms: "If you start going down the coal path or ore path, you eventually get headquarters to extract even more resources and make even more money. It's exponential" (ibid.). While the game's feedback loops could have been tuned to seek equilibrium, they were in fact designed to catalyze runaway collapse, which raises interesting questions about the degree to which the game is true to reality—whether it is scientifically "defensible" as Sheppard (2001) puts it, or how it combines a mix of fictional and nonfictional elements as Wells (2016) points out. The game's designers certainly intended to build a realistic model, as *SimCity*'s creative director Ocean Quigley tells: "We're doing our best to model real systems ... so that you'll understand something of how they actually work. And you'll make the tradeoffs that real cities have to make" (cited in Massey 2012, Mar. 12). The game's promotion of unsustainable decisions as a means to promote an understanding of sustainability as a complex system may be questioned on pedagogical, political, or moral grounds (see, for instance, Bereitschaft (2016), and Lauwaert (2007)). But it also raises broader questions about the underlying mechanisms by which *any* and *all* synoptic interactions represent their "source" worlds.

Procedurality and the Rules of Representation

It is not clear how aware Buckminster Fuller was to the inherent limitations of representing large complex systems when he suggested that his *World Game* would use a screen the size of a football field. The Argentinian author J.L. Borges, however, expressed the futility of perfect representation in his short fragment, *On Exactitude in Science*. It is worth reproducing here in full:

> ... In that Empire, the Art of Cartography attained such Perfection that the map of a single Province occupied the entirety of a City, and the map of the Empire, the entirety of a Province. In time, those Unconscionable Maps no longer satisfied, and the Cartographers Guilds struck a Map of the Empire

whose size was that of the Empire, and which coincided point for point with it. The following Generations, who were not so fond of the Study of Cartography as their Forebears had been, saw that that vast Map was Useless, and not without some Pitilessness was it, that they delivered it up to the Inclemencies of Sun and Winters. In the Deserts of the West, still today, there are Tattered Ruins of that Map, inhabited by Animals and Beggars; in all the Land there is no other Relic of the Disciplines of Geography. (Borges 1999, p. 385)

Borges's subtle irony aside, questions of representation (or when and how it becomes a facsimile) have been raised and debated in numerous studies of art, but they tend to focus on *content*, that is, on perspective, isometry, likeness, and so forth (for one important example, see Panofsky (1927/1997)). Interactive simulations such as *SimCity*, however, raise different concerns that stem from their particular *form*. Media theorist and game designer Ian Bogost (2007) captures the tension between reality and its representation by computational interactive media in what he calls "procedural rhetoric." It was briefly touched in the previous chapter as part of a discussion about the persuasiveness of interactive media vis-à-vis more traditional, text- or image-based media. The focus here, however, is not on rhetoric but on the other half of the concept, procedurality, or, more accurately, about how the two elements, rhetoric and procedurality, converge.

Generally speaking, procedurality denotes the sequential process by which computation operates. Computers execute rule-based calculations in sequence, as a way to represent diverse worldly phenomena digitally— as binary code (a series of ones and zeros). Binary code can then be read as the presence or absence of electricity in the computer's processing unit.[16] Once the world has been digitized, it can be reconstituted and made available to users through dedicated interfaces that range from more abstract symbol manipulation (programming, command line) to more intuitive graphic or embodied controls (graphic user interface or GUI).[17] Procedurality, therefore, implies a series of circular, responsive

---

[16] For a much more sophisticated account of computation, see Cantwell Smith (2002).

[17] Manovich (2001, p. 41) calls this "programmability" and argues that it is "the most fundamental quality of new media that has no historical precedent," although player pianos and, most famously, the Jacquard loom used "physically encoded instructions" (such as punch cards) as a means to program (and thus standardize) their operation (Sloman 2002, p. 197).

translations—from world to code to interface to code and so forth—that both represent the world (as digital material) and render it actionable (as a set of user interactions). As Bogost (2007, p. 4) writes,

> Software is composed of algorithms that model the way things behave. To write procedurally, one authors code that enforces rules to generate some kind of representation, rather than authoring the representation itself. Procedural systems generate behaviors based on rule-based models; they are machines capable of producing many outcomes, each conforming to the same overall guidelines. Procedurality is the principal value of the computer, which creates *meaning* through the interaction of algorithms. (emphasis added)

Bogost's introduction of meaning into the very core of procedurality reveals the process of computation to be both expressive and normative. It is expressive because, just as with any other form of media, computational authorship is selective. When programmers construct rule-based representations, they foreground, and thus express certain dimensions of reality and not others. At the same time, computation is also normative since software systems define a range of available and allowable, encouraged, or prohibited user actions that have clearly defined impacts on the system and, by extension, the world. In Bogost's (2007, pp. 45–46) words,

> Procedural representation models only some subset of a source system, in order to draw attention to that portion as the subject of the representation. Interactivity follows suit: the total number and credibility of user actions is not necessarily important; rather, the relevance of the interaction in the context of the representational goals of the system is paramount.

The convergence of expressivity with normativity illustrates the way interactive systems mobilize a particular interpretation of the world—how they promote particular dispositions not only in the way they represent or model a world but also in the way they enable paths of action on it. Procedural rhetoric, therefore, illustrates that for every interactive system, the coupling of the digital environment with the "real world" (or "source" environment) and the repertoire of available human-computer interactions are inseparable: representing the world and enabling action on it always go hand in hand. The implications are eminently visible in contemporary analyses of algorithmic culture and power (see, for instance, Bucher 2012; Kitchin and Dodge 2011; Pasquale 2015). But if we employ an ontological perspective, procedural rhetoric can also be seen as a generative phenomenon. Models not only describe and make actionable certain aspects of reality, but they sometimes bring that very reality to life, especially when we are dealing with complex systems that escape our mental models. Paul Edwards (2010, p. xiv), in this vein, argues that "Everything we know about the world's climate – past, present, and future – we know through models." And James Scott (1998) makes a similar case in regard to cadastral maps and statecraft: "a state cadastral map created to designate taxable property-holders does not merely describe a system of land tenure; it creates such a system through its ability to give its categories the force of law" (p. 3). Procedural rhetoric, in this sense, posits rule-based simulations (and synoptic interaction) as generative of that which they represent.[18]

Revisiting *SimCity* from the perspective of procedural rhetoric, we can first note that a simulation (or game) without bias is but a chimera; all simulations are forms of mediation, and all mediation is inherently discriminatory—echoing the boundary work that stands at the heart of complex systems thinking. But we can also see that the game's biases extend to both the way the game models complex urban systems and the range of actions it allows or even encourages users to take. If the meaning of the game emerges at the intersection of design intentions, system affordances, and user interpretations—as a co-created "narrative depicting the life of the digital city" (Wells 2016, p. 540)—then the game's promotion of unsustainable decisions spans all three (intentions, system, user): by weighing and rewarding certain behaviors disproportionately (promoting runaway feedback loops, for instance), and by excluding certain tools, options, and behaviors (such as the ability to create differential traffic restrictions or mixed-use zoning), the game conveys sustainability in a particular way.[19] This is not coincidental. As *SimCity*'s designers attest, the game aims to teach urban sustainability by promoting unsustainable decisions (encouraging mistakes is an effective learning tool according to Bergin's pedagogical patterns).[20] So as players accommodate the logic of the game's procedural rhetoric, they often end up making unsustainable choices despite their best intentions. When collapse ensues, or when the virtual

---

[18] Latour makes a similar point in the context of scientific inquiry when he suggests that Pasteur's discovery of microbes did not reveal a preexisting reality as much as generated it (Latour 1988).

[19] Zoning, in particular, was subject to several debates on *SimCity*'s sustainability (or lack thereof). See, for instance, Massey (2012, Mar. 12).

[20] See 'Mistake' in Bergin (2000, July).

city's dashboard screams of unsustainability, the game's pedagogical message comes through. In this sense, *SimCity*'s teaching of complex systems comes not through claims to faithfully representing reality (e.g., "the world is unsustainable") but by plunging players into the gaps between their own mental models of urban sustainability (e.g., "this is how a sustainable city should be designed") and the game's biased representation of urban design (e.g., "your actions will necessarily lead to unsustainability"). The game builds an understanding of urban sustainability by intermittently exposing and exploiting its own procedural rhetoric.

If the lesson of Borges's *Inexactitude* is that representation is unavoidably reductive, the lesson of Bogost's procedural rhetoric is that reductive representation has expressive and normative implications evident in both underlying model and interactive repertoire. We can conclude that computer simulations and the synoptic interactions that make them accessible encode a particular understanding of what the world is, how it can be grasped, and how it can (and should) be acted upon. We can see this more clearly by considering the application of synoptic interactions in an explicitly political context.

Not Your Everycity

So far the discussion of synoptic interactions focused on their capacity to help users make sense of complex systems. That capacity, I explained, depends not only on the degree to which an interactive system maintains fidelity to scientific data, but also on its procedural rhetoric: the way it interprets the world and renders it actionable for users as a repertoire of rule-based interactive affordances. This last aspect discloses the inherently normative dimensions of synoptic interactions—the manner by which they always represent reality in a way that corresponds with the social, political, economic, and other motivations of their creators—the way they embody what Feenberg (2017) calls the design code (as discussed in Chap. 1). While this may remain relatively benign when it comes to games that generalize city-building such as *SimCity*, it becomes much more acute when synoptic interactions are used in explicitly political contexts—when instead of modeling an "everycity," they represent a particular locale with its specific dynamics and tensions. In such cases the folding of epistemological and political dimensions, how we understand the complex system and how we may act on it, can become much more pronounced and, unsurprisingly, contentious.

Take, for example, MetroQuest (MQ), a planning decision-support tool that uses backcasting simulation and visualization techniques to engage the public with issues of urban sustainability design and policy-making.[21] Backcasting is the process of "envisioning desirable futures ... in order to stimulate discussions on how to get there" (Swart et al. 2004, p. 140). It allows users to generate and compare future urban development scenarios and then trace back the assumptions and conditions that led to those outcomes. While MQ can be used generically, much like *SimCity* only with a more pronounced focus on sustainability, it is more often created for a specific locale—a town, city, or region. During 2010–2011, a version of MQ (referred to here as MQ-V) was designed especially for the city of Vancouver, Canada. The design involved a group of stakeholders including members of the City of Vancouver's Transportation and Community Services departments, the Sustainability Office, and researchers from two of the city's universities (including myself). The new version was to be part of the City's public engagement on its incipient transportation plan ("Transportation 2040"), and in relation to the City's ambitions to become "the world's greenest city" by 2020 (Vancouver 2010). Consequently, MQ-V included design elements pertaining to land use (location and density of buildings and amenities), energy (the kind of energy produced and the location of energy facilities), and transportation (road allocation and designation of certain roads as high-capacity arteries). And it was to be used online, on mobile kiosks, and in facilitated workshops.

MQ-V features a challenge-based interactive flow (Ermi and Mäyrä 2005). It begins with a set of screens that explain the rationale behind the transportation plan, link the issues with the greater goal of making Vancouver green, and ask users to help the City decide on the degree to which it should pursue density, green transportation, and different energy production strategies. Introduction slides are followed by a set of 12 priorities that the user is asked to hierarchize. These are later used to communicate the effects of user choice: a green arrow indicates that the priority is served well by the user's choice, while a red arrow in the opposite direction means the inverse (see the right-hand side of the screen in Images 3.1

---

[21] The description that follows makes use of my previously published account of MetroQuest (Bendor 2012). For more on MetroQuest, see Carmichael et al. (2004), Haas Lyons et al. (2014), Robinson et al. (2006), Robinson et al. (2011), Rothman et al. (2002), Shapka et al. (2008), and VanWynsberghe et al. (2007).

Image 3.1   Bird's-eye view of transportation corridor and hub in MetroQuest. (Visuals by Nick Sinkewicz)

and 3.2). Once the list of priorities is put in order, the user can select one of three options for each of the three design elements (development, transportation, and energy). Every set of selections generates one of 27 possible scenarios, based on a scientifically defensible model and database first developed at the University of British Columbia (see in particular Rothman et al. 2002, p. 192).

For each scenario, MQ-V generates a set of three corresponding visualizations: a "community view" that features a wide-scope view of the transportation hub taken from a relatively high vantage point and which includes the adjacent neighborhoods and a detached, smaller-scale profile view of downtown (Image 3.1); a "street-level" view of a high-volume transportation corridor that includes a transportation hub (Image 3.2); and a "neighbourhood" view of detached homes, residential block located at the periphery of the transportation hub (using the same "street-level" visual perspective). Working within a particular scenario, users can switch between views, focus on a limited set of features ("highlights"), or get more information about the different urban design elements. Users can also switch between, explore, and compare any one of the 27 different

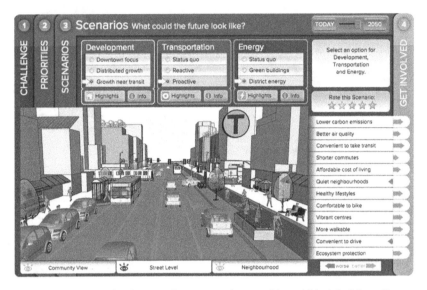

Image 3.2 Street-level view of transportation corridor and hub in MetroQuest. (Visuals by Nick Sinkewicz)

scenarios, telescoping back and forth within and between scenarios. After completing the iterative select-and-compare process, users are invited to rank scenarios on a scale of 1-to-5 stars, and see how their choices compare to others'.

Both MQ-V and *SimCity* are premised in the belief that spatiotemporal manipulation is key to uncovering the basic principles that underlie complex urban systems. But unlike *SimCity*, users of MQ-V do not have to wait for the consequences of their choices to become apparent several Sim-years later, thus bypassing the difficulties time delays pose for accurately evaluating a complex system (see, for instance, Meadows 2008, pp. 23–24; Meadows et al. 1972, pp. 144–145). With a push of a button or the sliding of a ruler, MQ-V provides immediate feedback and therefore encourages rapid experimentation and navigation within and between scenarios, visualizations, and timespans. Along with a simplified interactive repertoire that helps users focus on specific urban design features (and thus reduce the chance of "cognitive friction" (Cooper 2004)),[22] MQ-V's

---

[22] Complexity may produce positive "analytic uncertainty," but "simply layering on many systems and mechanics that interact with each other in complex ways makes it harder for players to grasp the system as a whole" (Costikyan 2013, p. 88).

immediacy makes it possible for users to discover the rules that underlie the system, to understand the relations between different drivers and variables, to identify potential sources of uncertainty, and thus develop appropriate mental models of the system as a whole (Robinson et al. 2006; Rothman et al. 2002). Furthermore, iterative scenario generation and comparison help MQ-V's users make connections between their own values and long-term goals, and the questions that inform public policy (Haas Lyons et al. 2014; Swart et al. 2004). Such a folding of descriptive and normative elements helps concretize links between individual and collective values, preferences, and positions, while introducing users to relevant policymaking terminology and allowing them "to actively explore the trade-offs and consequences associated with different policy choices, behaviours and objectives" (Robinson et al. 2011, p. 757). Users can see, for instance, how different transportation strategies result in different allocations of road space, in different sidewalk sizes, and different possibilities to regulate traffic inside residential neighborhoods. They can also witness these impacts on variable scales—from neighborhood to block to street—intermittently broadening and narrowing the scope of relevant policy questions, potentially increasing policy buy-in, and seeding social learning "which points to the inherent value of participatory processes in and of themselves as tools for social change" (ibid.).

As a way to clarify MQ-V's potential for social learning, and, by extension, its particular instantiation of epistemological and political factors, we can point out that MQ-V addresses not one but three separate, yet interrelated objectives. These roughly map onto what Andrew Stirling (2006) identifies as the three main justifications (or rationales) for public participation in environmental decision-making: normative, substantive, and instrumental justifications. Environmental communicators may recognize similar concerns in kindred models such as Daniels' and Walker's (1996) "collaborative learning," Senecah's (2004) "trinity of voice," and Walker's (2007) "participatory communication."

Normative justifications, writes Stirling (2006), reflect the belief that citizens have the right to influence the political processes that affect their lives. Public participation is posited as a virtuous end in and of itself—a way to build democratic skills, to advance social justice, and to form a countervailing political power that may keep government actions in check. When it comes to complex, wicked problems such as climate change or urban sustainability design, normative justifications gain added impetus from the fact that solutions often imply deep cultural, social, and economic transformations (Hulme 2009; Klein 2014). Leaving the public outside relevant decision-making processes not only flies in the face of democratic inclusivity but also makes public oversight of policy responses to wicked problems much more difficult. Furthermore, MQ-V reflects normative justifications not only by its raison d'être, but also by the ways it can be deployed. In Chicago, for instance, a similar (but much simplified) version of MQ was used during 2009 by the Chicago Metropolitan Agency for Planning as part of its public engagement on the region's GO TO 2040 plan.[23] In just three and a half months, it attracted 20,000 users and "resulted in clear public planning priorities that are reflected in the approved plan" (Haas Lyons et al. 2014, p. 24). With that said, we should keep in mind that the direct political impact of using tools such as MQ-V depends on questions of access (who gets to use them?), motivation (do users extend their learning to other areas of political activity?), and institutional responsiveness (are the entities that deploy the tool actually committed to integrating user preferences into their respective policies?). While using the tool may develop its users' sense of political self-efficacy, it may not necessarily lead to increased political influence.

Substantive justifications for public participation suggest that including a more diverse set of perspectives, values, skills, knowledges, and interests in processes of environmental decision-making will improve the quality of the decisions made. As Stirling (2006, p. 97) writes, "from a substantive point of view, participatory multi-criteria appraisal offers a means to be more rigorous about the questions that are asked of analysis, the way that they are addressed, the assumptions that are made in developing answers, and the interpretations and implications of results." To some extent, substantive justifications respond to the multi-scaled complexity of sustainability, in light of which sustainability issues cannot be solved by a single actor (Innes and Booher 2010; Talwar et al. 2011; van Kerkhoff and Lebel 2006). But they also reflect important challenges to the neat separation of "facts" from "values," as well as that between "experts" and "laypersons," as supported by contemporary approaches to scientific epistemology and practice.[24] We can

---

[23] See http://www.cmap.illinois.gov/about/2040 (last accessed Mar. 18, 2018).

[24] These approaches are often associated with social constructivism (Callon et al. 2009), "post-normal" science (Funtowicz and Ravetz 1993), and "mode 2" science (Nowotny et al. 2001).

see how MQ-V addresses substantive justifications in the way it seeks to build users' mental models of urban design and then provides them with a channel to compare and report their preferences. User choices can then become valuable inputs for policy, diversifying opinions and adding new insight.

Lastly, argues Stirling (2006), instrumental justifications value public participation for its raising awareness to the decision process, making it more accessible, transparent, and therefore legitimate. While this may reflect a genuine desire to build credibility, trust, and transparency, it may also reflect a desire to secure "implementation effectiveness" (Newig 2007), that is, promote policy "buy-in" by preempting the most vociferous forms of resistance to the suggested policy. In this sense, MQ-V can be seen as an attempt to navigate a complex policy terrain—to evoke at the same time a sense of detached impartiality and a sense of personal relevance. We can see this in the way MQ-V asks users about their individual priorities and then uses those priorities to display results in a meaningful manner. On the one hand, priority selection provides users with an opportunity to reflect on their own values and thus relate factual and normative dimensions. On the other hand, it helps to expose potential gaps between user values and user choices:

> Since the process of creating a future may involve conflicts, forcing users to make trade-offs, the policy decisions made while iterating through the series of time periods may or may not be consistent with the initially specific values. The degree of (in)consistency is reported to the user. Thus the user's performance is evaluated, not only for the achievement of goals, but also for consistency between their expressed values and the results of their policy choices. (Rothman et al. 2002, p. 185)

What *SimCity* achieves more cunningly—pushing players to make unsustainable choices, modeling runaway system dynamics, and frustrating players' initial designs for sustainable cities—MQ-V achieves through a series of more explicit prompts about values and preferences.

A similar dynamic unfolds through MQ-V's visualization strategy. As I explain in more detail in Bendor (2012), visual choices about perspective, location, scale, realism, and detail are all aspects of a delicate play between familiarity and impartiality. A few examples should suffice. Locating the user in an elevated, birdlike view—"the view of an absolute ruler" in Scott's (1998, p. 57) words—has the advantage of conjuring "many separate things at once," as Robert Harbison remarked on

Hollar's 1648 panoramic view of London (cited in Ford 2016). It is also less likely to trigger emotional responses to different scenario outcomes, while a street-level view that is closer to how the user experiences the city every day is much more relatable. In fact, the user's positioning in the second, "street-level" visualization (Image 3.2) had to be elevated from its initial position because users reported that the image evoked the feeling of sitting in traffic. In a similar vein, depicting a city block free of identifying markers helps to remove possible fears about the immediate, personal consequences of development (often expressed as NIMBY: "not in my back yard"), but visualizing several urban forms (a high-traffic corridor and a single home neighborhood) increases the chance that the user will find the scenarios relevant. On both accounts, policy options are animated simultaneously as personal and impersonal questions. I discuss the question of personal relevance in more detail in the next chapter.

The preceding discussion suggested that MQ-V addresses all three justifications for public participation in sustainability decision-making by offering an informed entry point into sustainability urban design, clarifying some of the latter's complex underlying questions, and opening up the space of sustainability policymaking for wider participation. It was also pointed out that MQ-V's particular way of relating epistemological and political dimensions foregrounds the often contradicting imperatives to make the complex system personally relatable while promoting a measure of impartiality—asking users to "think like a city." Yet, despite the designers' best intentions and evidence of its relative success in other jurisdictions (as briefly mentioned above), MQ-V was never deployed. City stakeholders were concerned that despite its use of a distantiating visual vocabulary, the tool would still evoke strong NIMBY responses. The City was still reeling from the woeful consequences of its previous sustainability urban design initiative, EcoDensity, and was not willing to risk a similar outcome.[25] We may surmise, then, that in this particular use of synoptic interactions, instrumental justifications trumped all others.

---

[25] See, for example, 'Vancouver EcoDensity hearing draws noisy turnout': http://www.cbc.ca/news/canada/british-columbia/vancouver-ecodensity-hearing-draws-noisy-turn-out-1.697412 (last accessed Mar. 18, 2018). The plan was rejected and ultimately cost the then-mayor his job.

FROM WORLD TO SELF AND BACK AGAIN?

Would Buckminster Fuller recognize his *World Game* in MQ-V's procedural rhetoric and, more generally, in synoptic interactions? To what extent does the imperative to "think like a city" consummate Fuller's dream? Had players of the *World Game* been able to experience the kind of impartiality implied and pursued by synoptic interactions, would they have been able to let go of their cultural presuppositions and personal interests or at least be able to recognize them for what they are? Fuller, it seems, was quite aware of the difficulties in negotiating the kind of knowledge necessary for engaging with complex systems. The *World Game* was designed to not only give players the opportunity to view the world system synoptically, but also to play out and test their own "theory" of how the world works. In this sense, the *World Game*, just like *SimCity* and MQ-V, shares the logic of synoptic interactions—the particular way in which synoptic interactions articulate the relations between epistemology and politics as a matter of navigating between multiple viewpoints and spatiotemporal positions and, no less important, seeking to balance individual and collective preferences. In this view, it is the addition of players and the integration of gameplay into the simulation that transforms the system from being merely complicated to being complex, emergent, or "wicked."

Although synoptic interactions do not always work successfully, the kind of effects designers pursue when they model complex systems and allow users to act on them merits critical reflection. I framed such anticipated effects by using Bogost's notion of procedural rhetoric, pointing to the fact that the way computational, interactive media model complex systems is only part of the picture. Procedurality implies not only that computational rule processing works in sequence, but also that such sequences invite user intervention through the system's interactivity. Modeling a complex system and providing users with an interactive repertoire with which to act on it are therefore two sides of the same dynamic: without appropriate modeling, the system's representation of the "source" world may remain vague and unappealing.[26] But without interactivity, the user may find it forbiddingly difficult to grasp the system's true emergent complexity. Tracing the ways in which the two combine

[26] Importantly, even well-modelled simulations often include a degree of randomness (Costikyan 2013).

through a system's procedural rhetoric, therefore, is key to understanding how synoptic interactions provide an entry point into the complex space that is sustainability, and the benefits and limitations of any such entry point.

The discussion of MQ-V's procedural rhetoric as a specific instance of synoptic interactions raises questions about the need, capacity, and benefits of balancing facts and values, "objective" and "subjective" elements. Does meaningful public participation in sustainability policymaking require the public to let go of their biases and disengage from lived experience as a form of justification, while adopting less personal, more impartial positions? And given the critique of fanciful rationalism in the previous chapter, is this even possible? From the perspective of consensus-seeking politics, the ability to disengage from personal interests (at least in principle) may be beneficial since consensus is improbable when participants refuse to critically evaluate their initial presuppositions and convictions, or are unwilling to explore alternative criteria for evaluating moral-political assertions. German philosopher Jürgen Habermas has made this view popular by his articulation of democratic politics as the capacity to facilitate and then safeguard rational discourse, that is, discourse characterized by informed exchange that does not appeal to some external criteria for truth (be it God, science, or authoritarian might). This form of (moral, political) discourse, Habermas (2001) argues, should not be satisfied with grounding truth in lived experience:

> Experiences support the truth claims of assertions; we maintain the claim as long as there are no dissonant experiences. But these truth claims can be redeemed only through argument. A claim grounded in experience enjoys provisional backing; as soon as it becomes problematic, we can see that a claim grounded in experiences is not yet by any means a justified claim. (p. 89)

There is clearly value in removing obstacles for consensual politics by letting the "unforced force of the better argument" shine through (ibid.), instead of endlessly debating the veridicality of one experience over another. The risks of allowing political discourse to be dominated by myth, conspiracy theories, or counterfactuals are all too well known.[27] But when Habermas asserts that "the success of deliberative politics depend not on

---

[27] Contemporary discussions about the impact of "fake news" are a revealing case in point (see, for instance, Milbank 2016, Nov. 18).

collectively acting citizenry but on the institutionalization of the corre-sponding procedures and conditions of communication" (1996, p. 27), we may wonder about the wisdom of reducing politics to its institutional grounding. It may be hard, in other words, to imagine transformative political processes taking place without passion, desire, or productive fric-tion—without emotional appeal and demonstrative argumentation.[28] Should we, or could we, bracket the vicissitudes of lived experience in pursuit of rational-critical discourse? Furthermore, if we accept that wicked problems raise perspectival problems *by definition*, and that Habermas's communicative action presupposes an initial agreement between actors "about definitions of the situation and prospective outcomes" (1990, p. 134), how likely is it that political dialogue and harmonization will pave the path for a better Anthropocene? Perhaps, as others have argued (Jacques Rancière, for instance), it may be precisely the drive for dissensus and not for consensus that reveals and potentially unravels the social, eco-nomic, and political entities that stand in the way of systemic change. This is, of course, a weighty question that cannot be fully addressed here. But it illustrates a key challenge for designers of synoptic interactions: is it pos-sible to create interactive structures that plot a path between impartial, consensus-driven trajectories and experiential, emotionally loaded, dissensus-provoking ones?

This leads us to another issue. This chapter discussed interactive media that respond to a particular view of sustainability as a complex problem to solve. As the problem reveals itself to be more and more complex, perhaps to the point of being intractable, our existing ways of dealing with com-plexity are showing their inherent limitations. Is it any wonder, then, that the cognitive expansion promised by computation—the emergence of the "macroscope," as de Rosnay (2011) calls it[29]—seems like the best way to regain our sense of control over our world? However, in Fuller's sugges-tion for a *World Game*, as in the other interactive systems discussed here,

---

[28] There are other important reasons to object to the Habermasian model of deliberative democracy, perhaps the most convincing of which are based on observations that there is nothing in dialogue itself that neutralizes power differentials (e.g., Hirschkop (2004), and the essays collected in Benhabib (1996)).

[29] "With its capacity for simulation, the computer has become a macroscope. It helps us understand complexity and act on it more effectively to build and manage the large systems of which we are the cells – companies, cities, economies, societies, ecosystems. Thanks to this macroscope, a new vision of the world is emerging, based on a unified approach to the self-organization and evolution of complex systems" (de Rosnay 2011, p. 305).

one may detect a wishful, optimistic conflation of politics with the logic of computation. It is quite nuanced: neither *SimCity*'s creators nor MQ-V's developers contend that their technology will "solve" sustainability for us. Nevertheless, despite the acknowledgment that sustainability raises normative and not only technical or epistemological conundrums, one may detect an underlying assumption that computation is indeed the solution for wicked problems—that throwing more computational power at the problems will eventually, and for Fuller ineluctably, lead to their disentanglement. As politics is substituted with complex calculuses of emissions and expenditures, computational simulations become the sword to sustainability's Gordian Knot. As I discussed above, such belief is blind to the way computation is always-already a proxy for human cognition and social power and thus merely transposes the problem to a different plain.[30] But pinning our hopes for a more sustainable world on the consequences of Moore's Law also seems to be the wrong lesson to learn from complex systems thinking. Donella Meadows (2008), prescient as ever:

> long before we were educated in rational analysis, we all dealt with complex systems. We are complex systems – our own bodies are magnificent examples of integrated, interconnected, self-maintaining complexity. Every person we encounter, every organization, every animal, garden, tree, and forest is a complex system. We have built up intuitively, without analysis, often without words, a practical understanding of how these systems work, and how to work with them. (p. 3)

Dealing with complex systems, it follows, is akin to remembering a more intuitive way to engage with the world; it calls for a kind of cultural anamnesis. Perhaps paradoxically, it asks us to learn a new modern vocabulary while demanding that we forget our inculcated, modern ways of seeing and relating to the world:

> People who are raised in the industrial world and who got enthused about systems thinking are likely to make a terrible mistake. They are likely to assume that here, in systems analysis, in interconnection and complication, in the power of the computer, here at last, is the key to prediction and control. This mistake is likely because the mind-set of the industrial world assumes that there is a key to prediction and control. (ibid., p. 166)

---

[30] For important discussions of the relations between computation and human cognition, see Weizenbaum (1976) and Dreyfus (1992). For analyses of computation as a form of social power, see Barry (2001).

Meadows's position may be vulnerable to accusations of romanticism, but if sustainability—a wicked problem that belies a host of other wicked problems like a set of nested Russian dolls—cannot be solved, that is, the kind of issues it raises cannot be brought under complete control, what are we to do? Meadows's answer is whimsical: "We can't control systems or figure them out. But we can dance with them!" (ibid., p. 170). But how? In *SimCity*'s insistence on promoting unsustainable decisions despite its users' original intentions, and in MQ-V's refusal to posit any one scenario as the ultimate solution, possible answers start to appear. Other possible answers that foreground the benefits of imaginative pluralism and playful ambiguity will be discussed in Chap. 5. But perhaps we should take Meadows's remarks as an invitation to explore alternative ways to re-conceptualize sustainability itself. In that spirit, the next chapter explores an approach that sees sustainability not as a problem to solve with the tools of rational analysis (be they holistic or not), but as a situation to work through.

BIBLIOGRAPHY

Adams, P. C. (1998). Teaching and Learning with SimCity 2000. *Journal of Geography, 97*(2), 47–55.

Barry, A. (2001). *Political Machines: Governing a Technological Society*. New Brunswick: Athlone Press.

Bendor, R. (2012). *Analytic and Deictic Approaches to the Design of Sustainability Decision-Making Tools* (pp. 215–222). Proceedings of iConference '12, Toronto.

Benhabib, S. (Ed.). (1996). *Democracy and Difference: Contesting the Boundaries of the Political*. Princeton: Princeton University Press.

Bereitschaft, B. (2016). Gods of the City? Reflecting on City Building Games as an Early Introduction to Urban Systems. *Journal of Geography, 115*(2), 51–60.

Bergin, J. (2000, July). *Fourteen Pedagogical Patterns*. Retrieved from http://csis.pace.edu/~bergin/PedPat1.3.html

Bogost, I. (2007). *Persuasive Games: The Expressive Power of Videogames*. Cambridge, MA: MIT Press.

Borges, J. L. (1999). *Collected Fictions* (trans: Hurley, A.). New York: Penguin.

Bucher, T. (2012). Want to Be on the Top? Algorithmic Power and the Threat of Invisibility on Facebook. *New Media & Society, 14*(7), 1164–1180.

Callon, M., Lascoumes, P., & Barthe, Y. (2009). *Acting in an Uncertain World: An Essay on Technical Democracy* (trans: Burchell, G.). Cambridge, MA: MIT Press.

Cantwell Smith, B. (2002). The Foundations of Computing. In M. Scheutz (Ed.), *Computationalism: New Directions* (pp. 23–58). Cambridge, MA: MIT press.

Capra, F. (1985). Criteria of Systems Thinking. *Futures, 17*(5), 475–478.

Carmichael, J., Tansey, J., & Robinson, J. (2004). An Integrated Assessment Modeling Tool. *Global Environmental Change, 14*, 171–183.

Connolly, W. E. (2006). Experience & Experiment. *Daedalus, 135*(3), 67–75.

Cooper, A. (2004). *The Inmates Are Running the Asylum*. Indianapolis: Sams.

Costikyan, G. (2013). *Uncertainty in Games*. Cambridge, MA: MIT Press.

Daniels, S. E., & Walker, G. B. (1996). Collaborative Learning: Improving Public Deliberation in Ecosystem-Based Management. *Environmental Impact Assessment Review, 16*(2), 71–102.

de Rosnay, J. (2011). Symbionomic Evolution: From Complexity and Systems Theory, to Chaos Theory and Coevolution. *World Futures, 67*(4–5), 304–315.

Devisch, O. (2008). Should Planners Start Playing Computer Games? Arguments from SimCity and Second Life. *Planning Theory & Practice, 9*(2), 209–226.

Dreyfus, H. L. (1992). *What Computers Still Can't Do: A Critique of Artificial Reason*. Cambridge, MA: MIT Press.

Dyball, R., & Newell, B. (2015). *Understanding Human Ecology: A Systems Approach to Sustainability*. London/New York: Routledge.

Edwards, P. N. (2010). *A Vast Machine: Computer Models, Climate Data, and the Politics of Global Warming*. Cambridge, MA: MIT Press.

Ermi, L., & Mäyrä, F. (2005). *Fundamental Components of the Gameplay Experience: Analysing Immersion*. Paper Presented at the 2005 DiGRA Conference: Changing Views – World in Play, Vancouver, Canada.

Espinosa, A., & Walker, J. (2011). *A Complexity Approach to Sustainability: Theory and Application*. London: Imperial College Press.

Feenberg, A. (2017). *Technosystem: The Social Life of Reason*. Cambridge, MA: Harvard University Press.

Ford, L. (2016). "Unlimiting the Bounds": The Panorama and the Balloon View. *The Public Domain Review*. Retrieved from http://publicdomainreview.org/2016/08/03/unlimiting-the-bounds-the-panorama-and-the-balloon-view

Forrester, J. W. (1998). Designing the Future. Lecture delivered December 15, 1998, at Universidad de Sevilla, Sevilla.

Fuller, R. B. (1969). *50 Years of the Design Science Revolution and the World Game: A Collection of Articles and Papers on Design*. Carbondale: World Resources Inventory, Southern Illinois University.

Fuller, R. B. (1999). *Your Private Sky* (edited by Joachim Krausse & Claude Lichtenstein). Zurich: Lars Muller Publishers.

Funtowicz, S. O., & Ravetz, J. R. (1993). Science for the Post-Normal Age. *Futures, 25*(7), 739–755.

Gaber, J. (2007). Simulating Planning – SimCity as a Pedagogical Tool. *Journal of Planning Education and Research, 27*(2), 113–121.

Goldstein, J. (1999). Emergence as a Construct: History and Issues. *Emergence,* *1*(1), 49–72.

Haas Lyons, S., Walsh, M., Aleman, E., & Robinson, J. (2014). Exploring Regional Futures: Lessons from Metropolitan Chicago's Online MetroQuest. *Technological Forecasting and Social Change, 82,* 23–33.

Habermas, J. (1990). *Moral Consciousness and Communicative Action* (trans: Lenhardt, C., & Nicholsen, S. W.). Cambridge, MA: MIT Press.

Habermas, J. (1996). Three Normative Models of Democracy. In S. Benhabib (Ed.), *Democracy and Difference: Contesting the Boundaries of the Political* (pp. 21–30). Princeton: Princeton University Press.

Habermas, J. (2001). Truth and Society: The Discursive Redemption of Factual Claims to Validity. In *On the Pragmatics of Social Interaction: Preliminary Studies in the Theory of Communicative Action* (pp. 85–103). Cambridge, MA: MIT Press.

Hirschkop, K. (2004). Justice and Drama: On Bakhtin as a Complement to Habermas. In N. Crossley & J. M. Roberts (Eds.), *After Habermas: New Perspectives on the Public Sphere* (pp. 49–66). Oxford/Malden: Blackwell Publishing/The Sociological Review.

Hulme, M. (2009). *Why We Disagree About Climate Change: Understanding Controversy, Inaction and Opportunity.* Cambridge/New York: Cambridge University Press.

Innes, J. E., & Booher, D. E. (2010). *Planning with Complexity: An Introduction to Collaborative Rationality for Public Policy.* London /New York: Routledge.

Johnson, S. (2001). *Emergence: The Connected Lives of Ants, Brains, Cities, and Software.* New York: Scribner.

Kim, M., & Shin, J. (2016). The Pedagogical Benefits of SimCity in Urban Geography Education. *Journal of Geography, 115*(2), 39–50.

Kitchin, R., & Dodge, M. (2011). *Code/Space: Software and Everyday Life.* Cambridge, MA: MIT Press.

Klein, N. (2014). *This Changes Everything: Capitalism vs. the Climate.* Toronto: Alfred A. Knopf.

Latour, B. (1988). *The Pasteurization of France* (trans: Sheridan, A., & Law, J.). Cambridge, MA: Harvard University Press.

Lauwaert, M. (2007). Challenge Everything? Construction Play in Will Wright's SimCity. *Games and Culture, 20*(3), 194–212.

Levin, S. (1999). *Fragile Dominion: Complexity and the Commons.* Cambridge, MA: Perseus Books.

Lilienfeld, R. (1975). Systems Theory as an Ideology. *Social Research, 42*(4), 637–660.

Lorince, J. (2013, March 6). Emergence (and Some Devastation) in Sim City. *Motivate. Play.* Retrieved from http://www.motivateplay.com/2013/03/emergence-and-some-devastation-in-sim-city/

Manjoo, F. (2013, March 4). The New SimCity Is Totally Addictive and Crazily Comprehensive. *Slate*. Retrieved from http://www.slate.com/articles/technology/technology/2013/03/simcity_review_the_new_version_of_the_classic_game_is_totally_addictive.html

Manovich, L. (2001). *The Language of New Media*. Cambridge, MA: MIT Press.

Massey, N. (2012, March 12). SimCity 2013 Players Will Face Tough Choices on Energy and Environment. *Scientific American*. Retrieved from https://www.scientificamerican.com/article/simcity-2013-players-face-tough-energy-environment-choices

Mayer, I. S. (2008). The Gaming of Policy and the Politics of Gaming: A Review. *Simulation & Gaming, 40*(6), 825–862.

McDermott, J. (2014, February 9). Using the New SimCity, 6 Urban Planners Battle for Bragging Rights. *Co.Design*. Retrieved from https://www.fastcoexist.com/1681515/using-the-new-sim-city-6-urban-planners-battle-for-bragging-rights

Meadows, D. H. (2008). *Thinking in Systems: A Primer*. White River Junction: Chelsea Green Publishing.

Meadows, D. H., Meadows, D. L., Randers, J., & Behrens, W. W., III. (1972). *The Limits to Growth; A Report for the Club of Rome's Project on the Predicament of Mankind*. New York: Universe Books.

Merchant, C. (2008). *Ecology* (2nd ed.). Amherst: Humanity Books.

Midgley, G. (2003). *Systems Thinking* (Vol. 4). London/Thousand Oaks: Sage.

Milbank, D. (2016, November 18). Trump's Fake-News Presidency. *The Washington Post*. Retrieved from https://www.washingtonpost.com/opinions/trumps-fake-news-presidency/2016/11/18/72cc7b14-ad96-11e6-977a-1030f822fc35_story.html

Newig, J. (2007). Does Public Participation in Environmental Decisions Lead to Improved Environmental Quality? *CCP (Communication, Cooperation, Participation. Research and Practice for a Sustainable Future), 1*, 51–71.

Nilsson, E. M., & Jakobsson, A. (2011). Simulated Sustainable Societies: Students' Reflections on Creating Future Cities in Computer Games. *Journal of Science Education and Technology, 20*(1), 33–50.

Nowotny, H., Scott, P., & Gibbons, M. (2001). *Re-thinking Science: Knowledge and the Public in the Age of Uncertainty*. Cambridge: Polity Press.

Odum, E. P. (1983). *Basic Ecology*. Philadelphia: Saunders College Pub.

Panofsky, E. (1927/1997). *Perspective as Symbolic Form*. New York: Zone Books.

Pasquale, F. (2015). *The Black Box Society: The Secret Algorithms That Control Money and Information*. Cambridge, MA: Harvard University Press.

Rees, W. E. (2012). Cities as Dissipative Structures: Global Change and the Vulnerability of Urban Civilization. In M. P. Weinstein & R. E. Turner (Eds.), *Sustainability Science: The Emerging Paradigm and the Urban Environment* (pp. 247–273). New York: Springer.

Rittel, H. W. J., & Webber, M. (1973). Dilemmas in a General Theory of Planning. *Policy Sciences, 4*, 155–169.

Robinson, J., & Cole, R. J. (2015). Theoretical Underpinnings of Regenerative Sustainability. *Building Research & Information, 43*(2), 133–143.

Robinson, J., Carmichael, J., VanWynsberghe, R., Tansey, J., Journeay, M., & Rogers, L. (2006). Sustainability as a Problem of Design: Interactive Science in the Georgia Basin. *The Integrated Assessment Journal, 6*(4), 165–192.

Robinson, J., Burch, S., Talwar, S., O'Shea, M., & Walsh, M. (2011). Envisioning Sustainability: Recent Progress in the Use of Participatory Backcasting Approaches for Sustainability Research. *Technological Forecasting and Social Change, 78*(5), 756–768.

Rothman, D. S., Robinson, J., & Biggs, D. (2002). Signs of Life: Linking Indicators and Models in the Context of QUEST. In H. Abaza & A. Baranzini (Eds.), *Implementing Sustainable Development, Integrated Assessment and Participatory Decision-Making Processes* (pp. 182–199). Cheltenham: Edward Elgar.

Scott, J. C. (1998). *Seeing Like a State: How Certain Schemes to Improve the Human Condition Have Failed*. New Haven/London: Yale University Press.

Scott, A. C. (2007). *The Nonlinear Universe: Chaos, Emergence, Life*. Berlin/Heidelberg: Springer.

Senecah, S. (2004). The Trinity of Voice: The Role of Practical Theory in Planning and Evaluating the Effectiveness of Environmental Participatory Processes. In S. P. Depoe, J. W. Delicath, & M.-F. A. Elsenbeer (Eds.), *Communication and Public Participation in Environmental Decision Making* (pp. 13–33). Albany: State University of New York Press.

Shapka, J. D., Law, D. M., & VanWynsberghe, R. (2008). Quest for Communicating Sustainability: Gb-Quest as a Learning Tool for Effecting Conceptual Change. *Local Environment, 13*(2), 107–127.

Sheppard, S. R. J. (2001). Guidance for Crystal Ball Gazers: Developing a Code of Ethics for Landscape Visualization. *Landscape and Urban Planning, 54*(1), 183–199.

Sisson, P. (2015, December 14). Check Out Buckminster Fuller's Simulation to Save the Planet. *Curbed*. Retrieved from http://www.curbed.com/2015/12/14/10621262/buckminster-fuller-world-games-columbia

Sloman, A. (2002). The Irrelevance of Turing Machines to Artificial Intelligence. In M. Scheutz (Ed.), *Computationalism: New Directions* (pp. 87–127). Cambridge, MA: MIT Press.

Stirling, A. (2006). Analysis, Participation and Power: Justification and Closure in Participatory Multi-Criteria Analysis. *Land Use Policy, 23*, 95–107.

Swart, R. J., Raskin, P., & Robinson, J. (2004). The Problem of the Future: Sustainability Science and Scenario Analysis. *Global Environmental Change, 14*, 137–146.

Talwar, S., Wiek, A., & Robinson, J. (2011). User Engagement in Sustainability Research. *Science and Public Policy, 38*(5), 379–390.

Taylor, P. J. (2005). *Unruly Complexity: Ecology, Interpretation, Engagement.* Chicago: University of Chicago Press.

Terzano, K., & Morckel, V. (2016). SimCity in the Community Planning Classroom: Effects on Student Knowledge, Interests, and Perceptions of the Discipline of Planning. *Journal of Planning Education and Research, 37*(1), 95–105.

van Kerkhoff, L., & Lebel, L. (2006). Linking Knowledge and Action for Sustainable Development. *Annual Review of Environment and Resources, 31,* 445–477.

Vancouver. (2010). *Vancouver 2020: A Bright Green Future.* Vancouver: City of Vancouver.

VanWynsberghe, R., Carmichal, J., & Khan, S. (2007). Conceptualizing Sustainability: Simulating Concrete Possibilities in an Imperfect World. *Local Environment, 12*(3), 279–293.

Walker, G. B. (2007). Public Participation as Participatory Communication in Environmental Policy Decision-Making: From Concepts to Structured Conversations. *Environmental Communication, 1*(1), 99–110.

Weizenbaum, J. (1976). *Computer Power and Human Reason: From Judgment to Calculation.* San Francisco: W. H. Freeman.

Wells, M. (2016). Deliberate Constructions of the Mind: Simulation Games as Fictional Models. *Games and Culture, 11*(5), 528–547.

Westley, F., Carpenter, S. R., Brock, W. A., Holling, C. S., & Gunderson, L. H. (2002). Why Systems of People and Nature Are Not Just Social and Ecological Systems. In L. H. Gunderson & C. S. Holling (Eds.), *Panarchy: Understanding Transformations in Human and Natural Systems.* Washington, DC: Island Press.

Woessner, M. (2015). Teaching with SimCity: Using Sophisticated Gaming Simulations to Teach Concepts in Introductory American Government. *PS: Political Science & Politics, 48*(2), 358–363.

Worster, D. (1994). *Nature's Economy: A History of Ecological Ideas* (2nd ed.). Cambridge/New York: Cambridge University Press.

CHAPTER 4

Experience

A Snowball Fight in Hell

Standing on the US Senate podium in February 2015, Republican James Inhofe of Oklahoma, the loudest voice in a chorus of climate change denialist lawmakers, reached into a ubiquitous, see-through Ziploc bag and pulled out a medium-sized snowball. Holding the snowball in his hand like it were a smoking gun, Inhofe exclaimed: "we keep hearing that 2014 has been the warmest year on record, I ask the chair, you know what this is? It's a snowball just from outside here. So it's very, very cold out. Very unseasonable."[1] Barely able to hold back a grin, Inhofe proceeded to toss the snowball in the direction of President Obama. "So, Mr. President, catch this!" (One could only hope the stunt did not catch the President by surprise.) Inhofe's theatrics aside, the tossed snowball raises interesting questions about the relations between our everyday experiences of the world and our ability to make sense of it and act appropriately. While Inhofe's gesture displays his (willful?) ignorance of both the difference between weather and climate and the links between climate change and extreme weather events such as unseasonal blizzards, drawing strong conclusions based on the vicissitudes of personal experience is not endemic to

---

[1] Video and transcript are available through *C-Span*'s website: https://www.c-span.org/video/?324568-2/us-senate-legislative-business&live=&start=6893 (last accessed Mar. 18, 2018).

© The Author(s) 2018
R. Bendor, *Interactive Media for Sustainability*,
Palgrave Studies in Media and Environmental Communication,
https://doi.org/10.1007/978-3-319-70383-1_4

Republican lawmakers. We all look to our experiences for evidence and, in this sense, are susceptible to overvaluing, undervaluing, misinterpreting, or simply failing to register certain experiences, especially if such experiences produce only "weak" perceptual signals (Bord et al. 2000, p. 206). The "boiling frog syndrome,"[2] in other words, is not exclusive to frogs.

Failing to perceptually register, or misinterpreting a phenomenon that has been perceptually registered, evidences a similar cognitive frustration: if we expect to experience a phenomenon in a particular way or with a particular intensity yet fail to do so, we tend to assume that the phenomenon has not taken place. In the case of climate change, since the media often points to rising temperatures as unequivocal symptoms of anthropogenic climate change, its absence—even if only temporary, and even if it could be explained away—may be taken as evidence of *The Great Global Warming Swindle*,[3] while its felt presence may compel individuals to believe in the reality of climate change and act upon it (Broomell et al. 2015; Egan and Mullin 2012; Li et al. 2011; Myers et al. 2013; Reser et al. 2014; Schuldt and Roh 2014; cf. Whitmarsh 2008). Seeing—or more accurately, experiencing—is believing.

The question that underlies this chapter, however, is not why we seek to validate knowledge in experience, or whether we stand on solid ground when we do so (a question that has occupied philosophers and scientists at least since Descartes famously questioned the trustworthiness of his senses).[4] Rather, I am interested here in the use of interactive media to create new experiences or reinterpret existing experiences in ways that may lead to forming new relations to sustainability—to bring sustainability "closer" and therefore more personally meaningful and significant. Occasionally, such a task may be achieved through a relatively simple act, the linguistic substitution of "global warming" with "climate change" is a

---

[2] In the context of climate change, the "boiling frog syndrome" was made famous by Al Gore's award-winning documentary, *An Inconvenient Truth* (2006). It is based on the now-discredited belief that if a frog is put into boiling water, it will immediately jump out to safety, while if it is put into tepid water that is increasingly brought to boil, it will remain in the pot until it dies.

[3] *The Great Global Warming Swindle* is the name of a documentary film that denies the reality of climate change. It was made by Martin Durkin in 2007.

[4] "I will suppose, then, that everything I see is spurious. I will believe that my memory tells me lies, and that none of the things that it reports ever happened. I have no senses. Body, shape, extension, movement and place are chimeras. So what remains true? Perhaps just the one fact that nothing is certain" (Descartes 2017, p. 20).

case in point (Weber 2016, p. 128; Whitmarsh 2009).[5] But the challenge remains significant since some of the most urgent sustainability-related dilemmas of our time do not necessarily lend themselves to immediate experience, either because they send only weak perceptive signals or because they have yet to take place. Where and how we choose to produce energy, the outcomes of industrial manufacturing and disposal, the loss of biodiversity, and the onset of climate change all indicate events that tend to occur outside of the majority of the public's everyday experience, remain unobtrusive, discrete, or time delayed, and therefore defy the sensorial linking of causes with effects (Cox and Pezzullo 2015; Moser and Dilling 2007, pp. 5–6; Tomlinson 2010; Weber 2016). It is simply impossible to reinterpret or reframe an experience that has not occurred.[6]

If we have not or maybe even cannot personally experience certain aspects of sustainability—at least not until it may be too late to make a difference—how can we develop the cognitive and social means necessary to de-abstract and engage with it? What may compel us to act with the kind of urgency and commitment needed to meet our compounding challenges? In Chap. 2, these questions were answered by way of skipping over understanding straight to action. Persuasive interactions were discussed there as a way to create sustainable attitudes and behavior regardless of whether people were fully aware of the assumptions that informed the behavioral technologies they used. In Chap. 3, however, answers focused on promoting complex systems thinking. With synoptic interactions, sustainability was to be known, "solved," or managed through spatial and temporal manipulation of the complex, emergent, intertwined subsystems associated with it. In this chapter, we shift our attention from reason to experience[7]: from communicative modalities that aim to build users' cognitive models of sustainability to communicative modalities that seek to evoke users' experiential affinities with sustainability. What I will call here

---

[5] George Lakoff's work on embodied metaphors and their ideological use (Lakoff 2004; Lakoff and Johnson 1980/2003) stands out in this context. There is, however, a price to be paid for linguistic substitution. Whereas "global warming" may evoke more concern than "climate change" (Whitmarsh 2009), it is easier to challenge based on individual experiences of weather (as Inhofe's antics demonstrate). This is perhaps why the former appears more often than the latter on websites run by conservative think tanks (Schuldt et al. 2011) and in Tweets originating from "red states" (Jang and Hart 2015).
[6] Bergson (2007) makes a similar observation: "it is impossible to travel back to an intuition one has not had" (p. 144).
[7] See Feenberg (2010).

resonant interactions foreground human experience as the key to perceiving and acting on sustainability. Such interactions seek to bring sustainability closer to our everyday perceptual world and thus make it more personally relevant. Resonant interactions, in other words, seek engagement with sustainability not by prompting the user to follow a behavioral script, nor by developing the user's capacity to master a complex system, but by conveying sustainability as a set of affective, emotional, resonant markers, a matter of felt significance and salience. How they do so is the subject of this chapter.

EXPERIENCE IN ENVIRONMENTAL COMMUNICATION

The claim made by resonant interactions is that lived experience is key to developing a certain kind of affinity with sustainability. But what is meant by "experience"? Is anything excluded from it? Are we not facing, in other words, a notorious catchall term, a "suitcase word" (Minsky 2006) into which many meanings could be packed, a proverbial rabbit hole? Given that experience, as Hans-Georg Gadamer remarked, is "one of the most obscure [concepts] we have" (cited in Jay 2005, p. 2), it is not my intention to provide a complete conceptual or genealogical account of the term, nor to provide a comprehensive analysis of its semiotic elasticity. Instead I wish to draw its contours with broad brushstrokes, allowing us first to relate the term to existing work in environmental communication and, second, to create a basis from which to understand the ways it is currently mobilized in the design of interactive media for sustainability.

My point of departure for discussing experience is what Martin Jay (2005) identifies as the term's inherent duality. On the one hand, experience references the visceral domain of sensation, perception, embodiment, and affect (*Erlebnis* in German, which I will call here "immediate experience"). This is what we mean when we say that we "had an experience." On the other hand, experience is also a marker of a more elongated biographical accumulation of events, the seat of memory, identity, tradition, and wisdom (*Erfahrung* in German, which I will call here "sedimented experience"). This is what we mean when we say that we "are experienced." If immediate experience connotes the ineffability and intensity of events undergone by individuals, the meaning of sedimented experience is much more socialized and therefore shareable— "the manifestation of the traditionary experience of a community"

(Arthos 2000, p. 3). If immediate experience is characterized as a pre-linguistic phenomenon, sedimented experience draws our attention to language, and the history, traditions, and cultural mores brought to bear by language on the raw materials of immediate experience. To borrow Agamben's (1993) distinction, the relation between immediate experience and sedimented experience is similar to that between voice and language, between pure expressivity and the ineluctable imposition of meaning.[8] Jay (2005, p. 6) reads experience (in more general terms) as "the nodal point" between immediate and sedimented experience, less a path for a veritable rapprochement between the two and more of an axis around which the different meanings of the term may be observed and organized. But perhaps the more important observation he makes is that in everyday use, experience ultimately signifies moments of newness or transformation:

> an experience, however we define it, cannot simply duplicate the prior reality of the one who undergoes it, leaving him or her precisely as before; *something must be altered, something new must happen, to make the term meaningful.* (ibid., p. 7; emphasis added)

It is this meaning of experience that will be carried forward here—the way it indicates the subject's capacity to undergo "a kind of surrender to or dependency on what it is not, a willingness to risk losing the safety of self-sufficiency and going on a perilous journey of discovery" (ibid., p. 405). Experience, in other words, denotes our capacity for change, transformation, for becoming other. It is like a journey: "you cannot know what you will discover on the journey, what you will do, what you will find, or what you find will do to you" (as James Baldwin wrote in a letter to Jay Acton (cited in Als 2017, Feb 13 and 20)).

The two notions of experience described above are evident in current work in environmental communication. Sedimented experience, for instance, can be seen as the premise for research that seeks a direct link between what people describe as significant life experiences (SLE), often in relation to role models, educational settings, and time spent outdoors,

---

[8] We can find an alternative classificatory system in Deleuze and Guattari's (1994) differentiation of affects, percepts, and concepts. Roughly described, affects mark the passing of material into sensation, percepts the registration of affects as distinct events, and concepts the linguistic categories with which we interpret percepts.

and their willingness to take environmental action later in life.[9] Louise Chawla (2006) summarizes the approach's main argument as follows:

> when children have access to the natural world, and family members encourage them to explore it and give it close attention, they have a strong basis for interest in the environment. To turn this interest into activism, they later need to build on this foundation through education, membership in organizations, or the careers that they pursue; but from their childhood experiences in nature through their own free play and in the company of significant adults, they carry the memory that the natural world is a place of such full and positive meaning that it justifies their most persistent efforts to protect it. (p. 76)

SLE research's failure to identify a "single all-potent experience that produces environmentally informed and active citizens, but many together" (Chawla 1998, p. 381) points directly to the kind of mnemonic stringing together of multiple immediate experiences which was identified above with sedimented experience. The latter, in this mode, is more a question of cultural sensibilities and influences than of moments with transcendental potentials.

Nevertheless, single moments of extraordinary prescience may indeed compel deep transformation. Canadian author J.B. MacKinnon's (2013) memory of returning home to find that the prairie landscape of his youth, "a place of magic" (p. 4), was now erased to make way for lawns and shopping centers is a good example. The contrast revealed by the experience, discovering that his childhood home has become his "lost Eden," he writes, "was the beginning of a journey that would change the way I see the natural world" (ibid., p. 5). Similar views on the power of immediate experience inspire research on individual responses to environmental risk. Elke Weber (2006), one of the leading researchers in the field, differentiates between what she calls (concrete) "experience-based" and (abstract) "description-based" perceptions of long-term environmental risk, noting that the latter lack in the kind of intensity needed to propel individuals to act with urgency. This leads her to suggest that "The absence of a visceral response on part of the public to the risks posed by global warming may be responsible for the arguably less than optimal allocation of personal and

---

[9] For recent examples of research into significant or formative life experiences as triggers for pro-environmental activity, see Reibelt et al. (2017), Stevenson et al. (2014), Wells and Lekies (2006), Williams and Chawla (2016).

collective resources to deal with this issue" (ibid., p. 103). Weber has since qualified her conclusions to account for the influence of worldviews and political ideology, gender, age, and nationality (Weber 2010, 2016), reflecting a much more complex and nuanced picture of the relations between "experiencing" and "believing."[10] But her suggestion that, "The concretization of future events and moving them closer in time and space seem to hold promise as interventions that will raise visceral concern" (2006, p. 116), echoes the rationale behind the media discussed below. Either way, immediate experience and sedimented experience should not be seen as entirely separate domains but as parts of the same mechanism by which we derive meaning from, and relate to, the world.

Whether we appeal to immediate experience or to sedimented experience, the need to de-abstract sustainability by moving it "closer in time and space" suggests that interactive media have a particular role to play as experiential proxies, that is, as means to create situations that cannot be experienced otherwise, whether due to physical limitations (events that happen in remote places) or temporal limitations (events that have taken place in the past or will take place in the future). Experiential proxies may also allow users to re-experience the same (or a similar) event but from a different perspective, thus imbuing immediate experience with a different set of cultural sensibilities and nuances. In this sense, immersive media may produce "teachable moments": events or contexts with transformative potentials (Hart and Leiserowitz 2009; Lawson and Flocke 2009), similar to that moment experienced and told by MacKinnon. As I suggest next, key to the functioning of interactive media as experiential proxies is an articulation of experience as a question of presence—presence in a time, place, or situation.

PRESENCE

Presence is a complex, multidimensional phenomenon. In what follows, I unpack it by drawing from three complementary perspectives: from the perspective of human geography, presence can be understood as the experience of place. From the perspective of interactive storytelling (or digital narratology), presence can be seen as the experience of agency. And lastly,

---

[10] See, for example, Hornsey et al. (2016) and Renn (2011). The influence of culture on the perception of risk has been influentially discussed by Douglas and Wildavsky (1982) and by Wynne (1992).

from the perspective of existential phenomenology, presence can be seen as an indication of ontological embeddedness. Taken together, the three perspectives provide us with a point of departure to consider presence in virtual environments, which is the most important characteristic of resonant interactions.

Placeness

Intuitively, presence is understood as the experience of being present *somewhere*. From a strictly physical perspective, this is equivalent to the experience of space, generally identified with volume. Human geographer Yi-Fu Tuan (1977) explains that the experience of space fundamentally involves movement which is, literally, an expansion in space: "Movements such as the simple ability to kick one's legs and stretch one's arms are basic to the awareness of space. *Space is experienced directly as having room in which to move*" (ibid., p. 12; emphasis added). Movement, of course, is managed by the sensorimotor system, but kinesthetic experiences of space also involve sight and touch. These allow us to sense and make sense of spacing and distance, size and shape. All in all, Tuan writes, "space can be variously experienced as the relative location of objects or places, as the distances and expanses that separate or link places, and – more abstractly – as the area defined by a network of places" (ibid.). Edward Relph (2008, p. 10) adds that perceptual space "is a space of action centred on immediate needs and practices." "Space," he continues, "is never empty but has content and substance that derive both from human intention and imagination and from the character of space" (ibid.). Our experience of space, it follows, goes far beyond physiological processes to include cultural, material, and psychological elements.

Consider, for instance, how maps communicate space and then feed into our very experiences of that space—how, in other words, they mediate processes of spatial awareness and proficiency (Kitchin and Freundschuh 2000). The recent discovery that using modern navigation technology (such as GPS or Google Maps) impacts the brain's navigational capacity provides further proof of this (Maxwell 2013, Mar. 8). Different stories, memories, or myths, not only technologies, may equally affect spatial experiences (Tuan 1977).[11] The way First Nations people understood the rainforest in what is now British Columbia is categorically different from

[11] This is playfully illustrated in Rebecca Solnit's (and colleagues) series of urban atlases.

the way European settlers and industrial loggers saw the very same trees (Vaillant 2005). Such situated, intersubjective formulations of space point to the way *space* can be transformed into *place* by social and cultural structures, how geographical nodes can become "centers of felt value" (Tuan 1977, p. 4). Importantly, the experience of place is altogether different from that of space since places are not just particular segments of a more abstract, continuous space. As Relph (2008) observes, "In our everyday lives places are not experienced as independent, clearly defined entities that can be described simply in terms of their location or appearance. Rather they are sensed in a chiaroscuro of setting, landscape, ritual, routine, other people, personal experiences, care and concern for home, and in the context of other places" (p. 29).

The transformation of a "given space" into a "particular place" (Thiel 1996, p. 138) implies the permeation of the former with symbolism, memory, history, and value (Cresswell 2015). Such a transformation may take various forms. It may involve a top-down imposition of order on abstract space by, for instance, naming, establishing a grid, or putting in place recognizable markers (Scott 1998; Tuan 1977, p. 136). Or it may involve more bottom-up practices, organic or spontaneous consolidations of habituated spatial practices such as walking home from work on a regular path, playing games in a particular park, shopping in the same store, or visiting regularly the same beach. In each of these cases, repeated behavioral patterns form around specific spatial sites, adding to the latter a sense of familiarity, meaning, and significance. Spatial practices that conjure placeness, however, can also form collectively. Drawing inspiration from the way Jane Jacobs, the famous urbanist, described her East Village street as a site for "an intricate sidewalk ballet," David Seamon (1979) suggests what he calls "place ballet" as a way to conceptualize the concretion of individual acts into collective place-making: "In place ballet, space becomes place through interpersonal, spatio-temporal sharing. Human parts create a larger place-whole. The meaning of the whole is normally expressed indirectly – through day-to-day meetings and an implicit sense of participation" (p. 59). When individuals habitually share a space through their activities, the space becomes a site of emotional attachment, value, pride, and care. To generalize, then, place encodes the transformation of space through experience, which is, importantly, a process by which self, community, and place co-emerge (Relph 2008, p. 34).[12] It is, as Moores

---

[12] Edward Casey calls this "constitutive coingredience" (cited in Seamon 2015, p. 391).

(2012, p. 104) puts it, an "experiential accomplishment," but one that brings the relationality of the world into focus: space becomes place through the embodied, enacted relations it anchors. Seen through the notion of placeness, presence is the experience of the meaningfulness of space, one that emerges in relation to individual and collective practices.

Immersion

Thinking about presence as an outcome of placeness, and not merely the occupation of space, allowed us to move beyond "pure" spatial properties and enter the realm of meaning and significance. Such a shift gains additional support by studies of interactive storytelling in which presence is often understood as immersion in a virtual world that unfolds narratively—the "projection of consciousness into the story world" (Mason 2013, p. 30). In more traditional media such as books, theater, and film, this projection takes place entirely in the reader's or viewer's imagination and requires them to actively suspend disbelief, that is, postpone the recognition that their physical world and the narrative world are not one and the same. In contrast, interactive, virtual environments invite users to actively create belief by giving them a quasi-physical presence in the story world. When users are allowed to inhabit and manipulate the virtual environment, the realness of the narrative world is reinforced *in action* (Murray 1997). Interactivity, then, is key to creating a sense of immersion.

In virtual worlds, interactivity may take the form of various embodied actions and reactions—swipe the pad and the character's head swirls; press a button and the character jumps; move a cursor and the character picks up an object. From the perspective of interactive storytelling, however, what matters is not only how "mechanical" forms of interactivity are qualitatively different from the navigational or interpretive strategies enacted by readers and viewers,[13] but how mechanical interactivity may pave the way to deeper forms of immersion. The latter, writes Janet Murray (1997),

... is a metaphorical term derived from the physical experience of being submerged in water. We seek the same feeling from a psychologically immersive experience that we do from a plunge in the ocean or swimming pool: the sensation of being surrounded by a completely other reality, as different as water is from air, that takes over all of our attention, our whole perceptual apparatus. (p. 98)

[13] See Murray (1997, p. 110), and Aarseth's (1997) notion of "ergodic literature."

Although the experience of immersion, like that of space, may evoke physiological sensations, the "flooding of the mind with sensation" in Murray's words (ibid., p. 99), it is strengthened when the user is provided with opportunities to act meaningfully in and on the virtual story. In this sense, immersion may be produced by embodied interactions much in the sense intended by Paul Dourish (2001), that is, when action ("what is done") is coupled to meaning ("what is meant").

Narrative immersion in virtual environments may take hold of the user when they enact or embody a character in an unfolding narrative, when they intentionally navigate virtual space, or when they act with or on virtual objects. Immersion is also dependent on the degree to which the user's expectations of the virtual environment are met and, a point to which I will return below, the extent to which the conventions that underlie the virtual world remain consistent (McMahan 2003, pp. 68–69). Although the diegetic richness of the virtual environment, that is, the detail, intricacy, and realism of the elements that make the virtual world, contributes to the environment's believability, it does not guarantee agency.[14] The latter, Murray argues, requires the ability to make choices among an expansive repertoire of available actions and to exert real effects on the virtual environment—to "encounter a world that is dynamically altered by our participation" (Murray 1997, p. 128). Stacy Mason (2013) adds that meaningful agency includes a sense of what is possible: "Achieving agency within interactive systems requires not only that we have control over certain aspects of the system, but that we understand the control we have, we know our limitations, and we are fluent with the [sic] our means of influencing the virtual space" (p. 41). We may conclude that when agency is enacted in well-crafted virtual environments, it can result in a powerful sense of immersion, of being present in an alternative world. And because presence in virtual environments is assimilated *as personal experience* (Murray 1997, p. 170), it may evoke strong, potentially transformative emotional responses.

Being-in

So far we have encountered relatively practical ways to understand presence, first as a property of inhabiting value-laden places and second as the enactment of meaningful agency over one's environment. In Heidegger's

---

[14] For our purposes, the diegetic refers to all elements of the narrative world (see also Tanenbaum 2014).

phenomenology, however, the practical dimensions of presence provide an entryway into a much deeper existential question.[15] Heidegger argues that considering presence as merely the occupation of space (think of a chair in a room, a house in the city, a person in a park, and so forth) fails to capture a more primary embeddedness in the world without which physical (or volumetric) presence is meaningless. Much like in the case of the complex systems discussed in the previous chapter, in Heidegger's analysis, the whole precedes its parts. We can glimpse this more fundamental embeddedness when instead of considering objects in a disinterested, contemplative, or "theoretical" way, we consider them in use (what Heidegger calls "concernful absorption"). Contemplating objects, Heidegger (1962) writes, may reveal their "presence-at-hand" (*Vorhandenheit*), but acting with them may reveal their "readiness-to-hand" (*Zuhandenheit*), which manifests the "kind of being which equipment possesses ... in its own right" (ibid., p. 98). This kind of focal shift, from observing an object to considering it *in use*, may therefore disclose an object's essence, the way it is embedded in a matrix of involvements, significances, and concerns— what Heidegger calls a "referential totality" (ibid., p. 99). Consider, for instance, the difference between merely contemplating a wrench and using it to fix a leaky tap. By observing the wrench, we may note its appearance and aesthetic qualities. We may perhaps be able to assume something about its material aspects and even speculate about what it can be used for and how. But when we use the wrench to fix the leak, the wrench, the tap, the sink, the materials they are made of, and the functions they serve in the larger context of human praxis (hygiene, domestic technology, hydrology, etc.) all come to the fore—materials, activities, and contexts blend into an expansive system of relations encoded in the thrust of activity. In this mode, the object in use manifests an entire "network metaphysics" (Feenberg 1999, p. 195).[16]

Human beings exhibit a similar quality insofar as they too are essentially situated and relational. This is why Heidegger articulates being as "Being-in-the-world," leading Don Ihde (1990) to remark that Heidegger's

---

[15] Importantly, presence plays a decisive role in Heidegger's thinking not as a noun but as a verb, *to presence* or *presenc-ing* (Malpas 2006). In this mode, presence conjures the dynamic alternation of background-foreground relations ("concealment" and "disclosure") that Harman (2007) identifies as the primary insight of Heidegger's philosophy.

[16] Heidegger would later describe this ontological effect with the phrase, "The thing things world" (1971b, p. 181).

account of being is relativistic, but not because it expresses some form of relativism, but because it is quintessentially relational. Relations, in other words, precede substance and existence (see also Ingold 2000, p. 149). This is also indicated by Heidegger's use of *Dasein*, literally meaning "being-there," to designate the human subject. To be human is to be situated, to always *Be-in*:

> From what we have been saying, it follows that Being-in is not a 'property' which Dasein sometimes has and sometimes does not have, and *without* which it could *be* just as well as it could with it. It is not the case that man 'is' and then has, by way of an extra, a relationship-of-Being towards the 'world' – a world with which he provides himself occasionally. Dasein is never 'proximally' an entity which is, so to speak, free from Being-in, but which sometimes has the inclination to take up a 'relationship' towards the world. Taking up relationships towards the world is possible only *because* Dasein, as Being-in-the-world, is as it is. (Heidegger 1962, p. 84; emphasis in origin)

The important conclusion here is that from an existential point of view, presence, "Being-in," or what Heidegger would later call "dwelling,"[17] both precedes and is enacted through our interactions with different worldly entities. Active participation, therefore, is "the most fundamental relation to reality" (Feenberg 1999, p. 196) not only because it reveals something about reality but also because it reveals something about the ontological openness of the human condition. Being-in signals the kind of experimental openness Jay locates as the very ground of experience. To be authentically present, then, is akin to becoming acutely aware of the relations that define us, while being ready to accept the possibility that those very relations may change us.

Considering presence from the triple perspectives of human geography, interactive storytelling, and existential phenomenology points to the tight relations between acting and experiencing. One may be present in a situation by simply experiencing it sensorially. But that experience may become even more meaningful and significant when it includes the capacity to act on the situation in meaningful, nontrivial ways. Such a capacity, it may be apparent, fits well the particular qualities of interactive media.

---

[17] See Heidegger (1971a).

How Does It Feel to Be a Tree?

The experience of presence in virtual environments is often triggered by embodied sensations. Minimally, this may entail providing users with navigational capacities that affect the virtual environment. Take, for example, *The Crystal Reef*, an interactive film that was produced by Stanford University's Virtual Human Interaction Lab.[18] The film is described by its co-director Lauren Knapp as "a 360-degree virtual reality science story" in which "Viewers are brought down underwater using virtual reality and presented with a visual metaphor of what the oceans will look like in a future affected by climate change."[19] Wearing VR (virtual reality) headgear, viewers are free to rotate the camera and focus on any aspect of the filmed environment, while listening to the narrator and an Italian ocean scientist discuss ocean acidification. The narrator explains that the film's use of VR technology is a means to generate a "teachable moment," providing actionable knowledge. In the narrator's words: "When people experience the future impact of climate change for themselves, alongside a scientific expert like Dr. Micheli, it becomes easier to understand the crisis."

The Crystal Reef demonstrates that adding a measure of interactivity to what is a compelling yet otherwise fairly standard science education documentary may increase the viewer's engagement. The ability to navigate a 360-degree stereoscopic screen effectively spatializes the film by adding width to the kind of depth viewers already enjoy with commercial 3D films. As result, the experience is much more captivating. But being able to *affect the action* on the screen is not nearly as powerful as *being part of the action* on the screen (Burdea and Coiffet 2003, p. 3). In other words, interactivity does not immediately translate to immersion. That, as discussed above, requires a heightened sense of agency, the feeling that the user is indeed part of the virtual world they inhabit. Such a feeling is more common in game environments, where being embedded in the events that unfold on the screen is a fundamental part of the experience.

In October 2016, game development company Ubisoft released its first VR game, *Eagle Flight*, for the Oculus Rift platform. Over the next two months, the game was also released for the PlayStation VR and HTC Vive platforms. Ubisoft introduces the game on its website as follows:

---

[18] *The Crystal Reef* can be viewed on Youtube, albeit with mouse-based navigation instead of full VR: https://www.youtube.com/watch?v=G0N7WFl6lbE (last accessed Mar. 18, 2018).

[19] See https://lauren-knapp.com/portfolio/the-crystal-reef (last accessed Mar. 18, 2018).

50 years after humans vanished from the face of the Earth, nature reclaimed the city of Paris, leaving a breathtaking playground. As an eagle, you soar past iconic landmarks, dive through narrow streets, and engage in heart-pounding aerial dog fights to protect your territory from opponents.

With innovative and intuitive controls, you quickly learn to perform impressive aerial maneuvers in the blink of an eye. Eagle Flight sends you to the skies to experience the freedom of flying and explore Paris from never-before seen perspective.[20]

Although gameplay includes such features as flying through floating rings, racing along obstacle-riddled pathways, capturing prey and bringing it back to the nest, and multiplayer aerial dogfights, it is the experience of flying that sets the game apart from other VR experiences.

The game requires players to wear a special headset, but once the player is wired, soaring through the skies requires only that the player gesture at the desired direction by moving their head. While the game also requires the use of a hand controller, the controller is used only minimally for accelerating, decelerating, and launching a high-pitched shriek that acts as a weapon. The game's mechanics are simple and quite intuitive. When I recently played the game, it wasn't long before I found myself gliding, rolling, and weaving freely through a gorgeously rendered Paris, its buildings and boulevards overgrown with vegetation, unencumbered by humans, and dotted with wildlife such as zebras and elephants(!). After less than an hour playing, I found myself in agreement with the superlatives heaped on the game: "Eagle Flight is quite simply, one of the most joyous experiences you can have in gaming" (Regan 2016, Nov. 22).[21] The game is truly exhilarating. But what makes it so?

The short answer is that the game provides a real sense of immersion. The premise is original, the visuals vivid, in-game objects realistic, and the sound (including the musical score) establishes the ambience perfectly.

[20] https://www.ubisoft.com/en-US/game/eagle-flight (last accessed Mar. 18, 2018).
[21] Naturally not all players feel the same. "Like so many other early virtual reality games, Eagle Flight feels less like a game and more like an experiment that was polished up and given a price tag," writes Chad Sapieha in his review for the *Financial Post* (Sapieha 2016, Nov. 9). I found Brian Albert's review for *IGN* to be both fair and accurate: "Eagle Flight surprised me with how quickly I felt at home flying and fighting with other birds above Paris. It features some of the best and most responsive and comfortable gameplay available on the PSVR, though like most current VR games its appeal may be short-lived if you're not a completionist who's crazy about collectibles or high scores. This is a weird idea, well executed, that soars high" (Albert 2016, Nov. 14).

But most importantly, the experience is truly engulfing: there are no "holes" in the virtual environment, and the motion feels so consistent and smooth that as I moved my head to set the eagle's flightpath, my body responded as if I was really airborne. This is not trivial. Early experiments with VR—from Glowflow (1969) to CAVE (1992)—required users to physically enter the virtual environment, that is, to step into a physical structure whose walls and sometimes floor served as surfaces for the projection of the virtual environment.[22] Interaction was full-bodied but required large specialized facilities.[23] The introduction of the data glove (commercially available in 1987) and the mounted headset (commercially available in 1994) made it possible to create more reasonably priced and practically sized VR experiences—a trend that continues to this day with cheap and simple devices such as Google Cardboard. Users could now experience extended haptic responsivity by engaging only a part of their sensorimotor system (head and hands). However, many early VR technologies were unable to successfully achieve "sensorial substitution" (when information that is usually derived from one sensorial domain is presented to the brain through another sensory system (Burdea and Coiffet 2003, p. 262)). This was often the consequence of insufficient computation power causing a lag between player movement and its reflection on the screen (motion mismatch). The lag prevents the player's physical body and their virtual body to function in unison. Frequent sensory conflict meant that until recently (and for some still), most VR experiences left users feeling motion sickness.[24] Full embodied immersion could only be achieved once this "doubling" of the user's body was solved, which, in the case of *Eagle Flight*, was partly achieved by limiting the field of vision when players make swooping head motions—a particularly nauseating maneuver.

The use of VR in *Eagle Flight* produces a direct kinesthetic sensation: to act *on* the game is to act *in* the game. As players learn to control the eagle's motion, the game's mediation dissipates, offering players something like

[22] For accessible, albeit early accounts of the history and social implication of VR, see Rheingold (1991), and Schroeder (1996).

[23] *Sensorama*, prototyped by Morton Heilig in 1962, presented a hybrid form: neither full-bodied nor entirely based on a mounted headset, it looked like an arcade game that invited users to stick their head into a small pod while gripping bicycle-like handlebars (see Rheingold 1991, p. 50).

[24] For more on the physiological issues involved in VR motion sickness, see Patterson et al. (2006).

"total transparency" (Ihde 1990, p. 93). Players become the eagle. But the strong physiological sensations produced by *Eagle Flight* are not the sole reason for the game's capacity to open "a magical window onto other worlds" (Rheingold 1991, p. 19). The magic is produced when the game manages to draw players into the virtual world by craftily combining sensorial triggers with a sense of agency. If we agree with Paul Dourish's (2001, p. 125) assertion that "Embodiment is a participative status, a way of being, rather than a physical property," then it is here that mere corporeality is overtaken by embodiment.

Dourish's definition of "embodied interactions" as something that qualitatively exceeds the boundaries of strict corporeality echoes Murray's argument that agency is not only a mechanical but a narratological condition. Creating rich immersive experiences requires more than merely dialing up embodied sensations, since immersive experiences depend on the way those sensations are made to resonate with the larger context, that is, how they couple action to meaning (Dourish 2001). In the case of *Eagle Flight*, this means evoking a powerful and consistent feeling of life as an eagle (or at least as imagined by game designers). It starts with the game's setting, a post-humanity Paris that eerily echoes the kind of speculative reality described in Weisman's provocative *The World Without Us* (2007). The game's narrator, exuding Attenboroughesque *gravitas*, adds a measure of realism to the fictional world and colors the game with the kind of sensibilities we would normally expect from a *National Geographic* documentary, not a videogame. But most importantly, players experience the game exclusively from the perspective of the eagle, and while, on the one hand, this may be seen as a gimmick, merely an excuse for flying in VR, on the other hand, it produces a viscerally non-anthropomorphizing experience. The game, in other words, provides players with a resonant experience of what we may call "lived eagleness." And as the humanless streets of Paris become both familiar and alienating, ominous yet inexplicably inviting, the game's physical *space* transforms into a lived *place*, but one lived by, and for, eagles, not humans.

Providing users with an otherworldly experience is also the aim of *Tree*, a hybrid VR-tangible interactive film that "transforms you into a rainforest tree."[25] More an interactive installation than a film, *Tree* is one of several

---

[25] https://www.treeofficial.com (last accessed Mar. 18, 2018). The film was created by Milica Zec and Winslow Porter, and produced in collaboration with Rainforest Alliance.

recent attempts to bring the public viscerally closer to the natural world,[26] and was featured in both Sundance and Tribeca film festivals. Sundance Festival's website describes it as follows:

> See and feel what it's like to be one of the most essential building blocks of carbon-based life: a tree. This haptically enhanced virtual reality experience allows you to become a seed that is planted and grows into a full-sized tree in a majestic rain forest. With your arms as the branches and your body as the trunk, you experience the tree's growth from a seedling to its fullest form, taking on its role in the forest and witnessing its fate firsthand.[27]

Tree combines touch, sound, vision, and scent into a unique multisensorial experience. Users are equipped with a Subpac vibrating backpack unit, EMS (electronic muscle stimulation) sleeves, earphones, and an Oculus Rift VR viewer. Before entering the virtual space, users are asked to plant a real kapok seed in soil, creating a tangible precursor to what they are about to experience. The smell of soil will remain present throughout the experience, providing users with a continuous bridge from the virtual world to the real one. The rainforest is rendered beautifully, with insects, birds, and monkeys giving the rainforest a vibrant, teeming ambience. Wildlife interacts with users through the EMS technology. Birds may land on the user's hands, resulting in tangible sensations. The film also makes use of wind, and heat accentuates the experience's dramatic climax: a forest fire that ends the tree's life. As the tree falls to the rainforest's floor, the ground beneath the players vibrates to complete the experience.

Users may not be able to significantly impact the tree's surrounding or affect the tree's fate, and in this sense, the experience is perhaps closer to *The Crystal Reef* than to *Eagle Flight*. Nonetheless, those who experienced *Tree* report strong emotional responses. The installation's co-director Winslow Porter's testimony that "Many people cried or shouted during the final moments of Tree" is echoed across several reviews of the installation (Jaffe 2017, Feb. 23). In his review for *The Verge*, Adi Robertson (2017, Jan. 26) writes: "Tree takes the concept of virtual reality empathy to (literally) new heights," while *TechCrunch*'s reviewer, Josh Constine (2017, Jan. 20) writes: "I watched one woman come out of VR crying, having so fully

---

[26] See, for instance, *Treehugger: Wawona*, which, like *Tree*, was included in the 2017 edition of Tribeca Film Festival.

[27] http://www.sundance.org/projects/tree-79496ca6-8d2d-4913-b497-354e323318f3 (last accessed Mar. 18, 2018).

identified with the now fallen tree." *Tree* may not provide participants much in terms of agency, but its "perceptual realism" (McMahan 2003, p. 75) is certainly powerful enough to evoke strong emotional responses. The resonant proof is in the virtual pudding, so to speak. In this sense, if we take seriously Maurice Merleau-Ponty's observation that "we are in the world through our body" (1962, p. 239), *Tree* provides a vivid example of how resonant interactions may impact both what we take as "we" and how we consider "the world." Resonant interactions, therefore, illustrate promising ways to trigger a deep re-evaluation of what human and nonhuman life in the Anthropocene may look and feel like.

TIME TRAVELING IN VR

One of the striking elements in *Eagle Flight*, the VR game discussed above, is the virtual environment it illustrates. It is a futuristic world in which the urban landscape bears the traces of a discontinued modern civilization. This kind of movement in time, the ability to experience past or future worlds, is one of the most interesting capabilities of VR and a potent form of communication and learning (Otto and Wilkinson 2012). It allows environmental communicators and media designers to bring to life various realistic or speculative futures, turning scientific representations of future scenarios into spaces of semiosis. They allow, as the futurist Stuart Candy (2010) writes, the creation of "experiential futures." Such futures may serve different purposes and therefore be constructed with different virtual building blocks. Like the simulations discussed in the previous chapter, they may remain committed to some form of scientific defensibility (Sheppard 2005),[28] or exercise a larger degree of creative freedom, seeking believability instead. In both cases, the strong embodied elements that were the focus of the previous section are treated here as mere devices for telling a story about the future. In this sense, they function both diegetically and mimetically.[29]

Take, for example, Owl, a VR tool that was used in 2013 to engage the public on proposals for the redevelopment of San Francisco's Market Street, part of the Better Market Street initiative.[30] Using Autodesk's

---

[28] Sheppard (2005) sees defensibility as a measure of the transparency by which a visualization can be traced back to its scientific or logical underpinnings.

[29] According to the principles of Greek drama, the mimetic shows while the diegetic tells.

[30] http://www.bettermarketstreetsf.org (last accessed Mar. 18, 2018).

Infraworks modeling software and an attached iPad's compass and gyro-scope, Owl transforms 2D images into a panoramic, responsive 3D view. Peering into Owl's binocular-like viewfinder—an apt metaphor for the kind of time traveling the device offers—members of the public were able to switch between three different development proposals and experience them as immersive environments. This forms a tangible bridge between the assumptions and language of sustainability policymaking and everyday life (Gordon et al. 2011; Jégou and Gouache 2015; Reser et al. 2011; Sheppard 2005).

As discussed above, to be conceptually aware of environmental phe-nomena such as climate change is not the same as to experience them: "It's like 'hearing' Beethoven's fifth symphony by looking at the score. Unless you're a trained musician it's little more than lines and squiggles, not the majestic music that results in the execution of the symbolic repre-sentation" (Schueneman 2013, July 26). Owl fills this experiential gap with a much more tangible, visceral impression of environmental futures. This allows users to contrast past and present environments, compare visions of one or more futures, or simply view firsthand the possible impacts of a variety of sociotechnical and environmental dynamics. Research that examined the outcomes of a later application of Owl to communicate the flood risks implicated with climate change in Marin County, California, confirms the value of making tangible and making personal:

> Can 3D visualization of climate change impacts and adaptation options – using the OWL technology – increase viewers' level of concern, and the desire to engage in the local adaptation process? The overarching answer is a resounding yes … 75% of the population that came in with low to no con-cern [about existing flooding risks and increased risks due to sea level rise] indicated that they felt more concerned after viewing the future sea level rise scenarios.… The largest number of respondents (29%) indicated they would like more information about the process, another 19% said they would attend a meeting, and the smallest number of respondents (13%) said, they would like to take an active role in their community. The latter two groups combined (32%) constitute a surprising large proportion of community members who could be mobilized for civic engagement. (Moser et al. 2016, pp. 33, 34, 36)

Resonant interactions, we may say, can help the public de-abstract envi-ronmental phenomena and make them much more personally relevant, producing virtually, as it were, the same kind of snowballs tossed on the floor of the US Senate.

Because Owl is often used in policymaking contexts, its capacity to create resonant interactions walks a fine line in terms of its reliance on scientific defensibility. Although the visualizations do not depict "real" events or situations, they are neither entirely imaginary nor completely speculative. Yet, as a tool for formal public engagement (in both San Francisco's Market Street and in Marin County), the future possibilities displayed by Owl cannot seem fictitious. Owl's quasi-realness, we may say, materializes a firm boundary around what it can and cannot show and as consequence limits its use. Other immersive tools, however, especially those used outside formal policymaking processes, may be less obliged to comport with scientific defensibility—as long as they maintain a degree of believability, which is, in this case, a product of the immersive environment's capacity to conjure a world whose different elements fit together consistently. Believability, in other words, is not a measure of scientific defensibility but an outcome of in-world or diegetic consistency.

A brief examination of a "guerilla art" installation that was staged in Vancouver's English Bay during November 2014 may be instructive. The installation was commissioned by the Canadian Environmental group, Dogwood Initiative, and created by Rethink Marketing in collaboration with Adrian Crook and Associates (AC+A), and Factory 1 Studios. The installation was part of Dogwood's #NoTankers campaign, which responded to efforts by large multinational corporations (chief among them is Texas-based Kinder Morgan) to attain approval for expanding the Trans Mountain Pipeline and thus nearly triple the volume of diluted bitumen carried from Alberta's tar sands to British Columbia's coast from 300,000 to 890,000 barrels a day.[31] The bitumen, a dangerous substance which is extremely difficult to contain (Sassoon 2012, Aug. 20), would then be loaded on oil tankers that would carry it to be refined and sold in Asian markets. All in all, it was expected that the newly expanded pipeline would drive the number of oil tankers that currently visit the Vancouver area from 60 to about 400 annually.

---

[31] For more on the pipeline, see https://www.kindermorgan.com/business/canada/transmountain.aspx (last accessed Mar. 18, 2018). On the campaign, see https://dogwoodbc.ca/campaigns/no-tankers (last accessed Mar. 18, 2018).

Image 4.1   Oil spill in Vancouver's English Bay, part of a public VR installation commissioned by Dogwood Initiative. (Image by Lasko Konopa, Factory 1 Studios)

The installation comprised a single pair of responsive binoculars, shaped much like the Owl, installed with the Oculus Rift VR viewer, and placed on a path overlooking the beach. Peering through the binoculars, passersby could witness a horrific scene unfolding in 3D: a tanker sinking in the middle of the beautiful bay while emergency response teams try to contain the spill, smoke billowing from several fires, oil-slicked water, tar covering the shoreline, and a beached orca whale (see Image 4.1). The 3D panoramic image was animated and responsive: as users panned the binoculars, the image adjusted accordingly. The result was immersive, engaging, and unsettling: "the first person to see the 'catastrophe' actually swore loudly and did a double-take."[32] The double-take, which could be clearly seen in the video that accompanies AC+A's technical report, should not be discounted for it evidences the extent to which the situation revealed by the binoculars was indeed believable.

Unlike the immersive panoramas viewed through the Owl, the installation had no pedagogical aims. It was not meant to prop knowledge about

[32] http://adriancrook.com/our-work/oculus-rift-oil-spill (last accessed Mar. 18, 2018).

the dynamics and consequences of an oil spill but to stir an emotional response—to make it more personal. Kai Nagata, former Energy and Democracy Director with Dogwood Initiative, made that clear: "A picture is worth a thousand words ... And a 3-D virtual reality tour is worth a few more than that.... We wanted to bring it home, and give people a little flavour of what this could look like here in our city" (cited in Ball 2014, Nov. 12). To help make the oil spill feel personal, the installation drew a stark contrast between Burrard Inlet's natural beauty and the catastrophe revealed through the binoculars, a contrast made especially potent when encountered at the very site of the imaginary oil spill. Vancouver residents that take pride and joy in the city's natural beauty could not remain indifferent to the images. For them, individual and collective experiences of English Bay enriched the space depicted by the installation with layers of memories and meaning. In a playful reversal of spectacularity (Debord 1994), place fed back into space, closing the distance between spectator and spectacle. A future represented became a future lived, and with it the absurdity of Kinder Morgan's claims that an oil spill would have some positive effects—"Spill response and cleanup creates business and employment opportunities for affected communities, regions, and cleanup service providers" (O'Neil 2014, Apr. 30)—could not be denied. It seems quite ludicrous to entertain economic arguments for oil spills while in the presence of an orca whale lying dead on your favorite beach.

Every Traveler Needs a "Mattering Map"

This chapter illustrated resonant interactions, a form of technologically mediated interactions that seek to bring sustainability closer to our everyday perceptual world and thus make it more personally relevant, meaningful, and salient. It is for this reason that resonant interactions conjure a sense of presence, allowing the user to inhabit sustainable or unsustainable situations and develop a felt relation to the world. The way resonant interactions promote emotional investment in issues pertaining to sustainability reflects a view of the latter as a situation to inhabit or an event to undergo. Sustainability is materialized as felt embeddedness with transformative potentials.

Acting as experiential proxies, media that make use of resonant interactions can help communicate sustainability-related phenomena that register only "weak" perceptive signals (Bord et al. 2000). They may help de-abstract phenomena that are otherwise largely represented with scientific data, providing a more robust epistemological and motivational foundation for active engagement with sustainability. For instance, resonant interactions may help overcome what *The Guardian* has recently termed the "climate paradox" by establishing "a conductive connection between the high-resolution picture of the future that's been painted by intensely reviewed and re-reviewed scientific inquiry, and our responsibility to future inhabitants of the atmospheric stew into which we're injecting gigatonnes of greenhouse gases" (Joshi 2017, June 7). Resonant interactions, in this sense, can "connect numbers to feelings," or, as the IPCC's Working Group 3 asserts, "people's experience can make climate a more salient issue" (IPCC 2014, p. 164). Resonant interactions may not be capable of reversing or mitigating the effects of nature-deficit disorder (Fletcher 2017) and, despite *Tree*'s success, may not be able to truly bring to life wilderness in a digital form (Stinson 2017). But resonant interactions can create bridges between sustainability policymaking and everyday life, translating impenetrable technical language into the idioms of felt experience. As Gordon et al. (2011) suggest, "the more ways a process can be immersive, the more effective that process can be at engaging the public in discussing planning decisions" (p. 507).

With that said, a recent commentary in *Nature Climate Change* (Chapman et al. 2017) cautions against simplifying and overvaluing the efficacy of emotionally resonant messages. The commentary was written in response to David Wallace-Wells's unsettling article in *New York* magazine, "The Uninhabitable Earth" (Wallace-Wells 2017, July 9). In it, Wallace-Wells provides a detailed description of an almost unimaginable future world, giving credence to his warning that "no matter how well-informed you are, you are surely not alarmed enough" with the assurance that his account is not a work of science fiction:

> This article is the result of dozens of interviews and exchanges with climatologists and researchers in related fields and reflects hundreds of scientific papers on the subject of climate change. What follows is not a series of predictions of what will happen – that will be determined in large part by the much-less-certain science of human response. Instead, it is a portrait of our best understanding of where the planet is heading absent aggressive action. (ibid.)

Where the planet is heading "absent aggressive action" horrified many. And while Wallace-Wells's "doom and gloom" account was not strictly apocalyptic—it assumed humans would find ways to adapt to their new climatic reality—the vividness of his descriptions, his willingness to say the "unsayable," and his careful linking of climate change's ecological, social, and political dimensions made the article both fascinating and horrifying (much like watching a car accident, except in this case the car is being driven by the reader). The article became an overnight sensation and attracted praises and criticisms from some of the world's best-known climate scientists.[33] The commentary in *Nature Climate Science*, however, sought not to challenge Wallace-Wells's analysis but to problematize some of the assumptions made about the story's impact or, more specifically, about the perceived utility of using emotional messaging as a stand-alone communicative strategy:

> Rather than treat emotion as a lever or switch to be directly calibrated and pulled for a desired effect, the climate change communication community should adopt a more nuanced, evidence-based understanding of the multiple and sometimes counterintuitive ways that emotion, communication and issue engagement are intertwined. Emotions should be viewed as one element of a broader, authentic communication strategy rather than as a magic bullet designed to trigger one response or another. (Chapman et al. 2017, p. 850)

Translated into the language used in this chapter, it seems that leaning heavily on resonant interactions alone may not yield the intended transformative results. The public needs a way to digest and integrate emotional messages into more extensive mental, cultural, and ideological frameworks (Egan and Mullin 2017).

As argued above, resonant interactions rely on embodiment to conjure a sense of presence in another environment. In the context of interactive media, such presence is often understood as a form of immersion. It is induced by a corporeal sense of presence or, more precisely, by the overcoming of the corporeal doubling—having a body in the immersive environment and outside of it. Yet, resonant interactions do not end with corporeality. Captivating immersive experiences layer spatial experiences with the kind of meaning and significance we associate with placeness, and make use of the interactive affordances built into responsive, virtual environments to create a tangible sense of agency. The ability to act meaningfully on and through the virtual environment amplifies the user's sense of

---

[33] See http://nymag.com/nymag/letters/comments-2017-07-24 (last accessed Mar. 18, 2018).

presence. With that said, we should keep in mind that agency *within* the virtual environment should not be confused with the agency wielded by the virtual environment's authors (or designers). As Murray (1997) clarifies:

> There is a distinction between playing a creative role within an authored environment and having authorship of the environment itself.… interactors can only act within the possibilities that have been established by the writing and programming. They may build simulated cities, try out combat strategies, trace a unique path through a labyrinthine web, or even prevent a murder, but unless the imaginary world is nothing more than a costume trunk of empty avatars, all of the interactor's possible performances will have been called into being by the originating author. (p. 152)

Unless the virtual environment supports true emergent behavior, agency in virtual worlds will remain bounded by what the designer deems necessary. Computer code, in other words, is law (Lessig 2006).

Agency aside, John McCarthy and Peter Wright (2004) suggest that a holistic approach to mediated experiences should take into account all four "threads" of experience: the compositional (the fit of different diegetic element), the sensual (the feeling generated by the texture and ambience), the emotional (the coloring of experience by moods), and the spatiotemporal (the location of the experience in a particular place and time). When all four are woven together successfully, the user may genuinely feel transported to a different environment. We may add that at the "shallow" end of the experiential spectrum, users may merely partake in well-scripted mechanical interactivity, acting as conduits for the continuous flow of motion into emotion. On the "deeper" end of the spectrum, users may glimpse their essential embeddedness in the world, their fundamental relationality or *Being-in*, in Heidegger's terms (Bendor 2017). It is here that our ontological openness to the world—our willingness to go on a "perilous journey of discovery" as Jay writes—can be seen as an exciting opportunity for designing mediated, transformative experiences. The current popularity of VR as a means for social communication is indicative of this.[34] The temptation to fetishize virtual reality experiences, however,

---

[34] For recent examples, see *The Enemy*, a VR experience created by war photographer Karim Ben Khelifa (http://theenemyishere.org; last accessed Mar. 18, 2018), and the Academy Award-winning VR installation, *Carne y Arena*, by film director Alejandro González Iñárritu (http://carneyarenatlatelolco.com; last accessed Mar. 11, 2018).

must be acknowledged. Transcendental experiences are neither inevitable nor can they be conjured simply on demand. One may experience life as a tree without it leading to a heightened sense of environmental awareness, just as one may experience a bird's perspective without understanding it as a symptom of humanity's civilizational failure. In other words, there is nothing inherent to resonant interactions that guarantees that they will evoke the kind of nearness discussed above. This is because how resonant interactions affect users has much to do with the way they relate diegetic (in-world) elements to the extra-diegetic contexts they wish to communicate. It is here that we can see how in the "real world" of perception, immediate experience (*Erlebnis*) and sedimented experience (*Erfahrung*) are mutually dependent.[35] New sensory triggers, affects, and percepts wash over past experiences and give them new significance, while past experiences, memories, and social relations reframe and "slot" visceral impressions into a more elongated, meaning-giving contexture. In the experiential intermingling of the "now" and the "then," neither is left unchanged (Dewey 1934, p. 272).

"The great traveler," writes Walter Benjamin (1929/1999), "passes through cities and countries with anamnesis; and because everything seems closer to everything else, and hence to him, since he is in their midst, all his senses respond to every nuance as truth" (p. 248). The unadulterated receptivity Benjamin describes has undeniable phenomenological tones, but truth, as Richard Rorty (2007) argues persuasively, is intersubjective and therefore not immune to social effects. The fundamental openness marked by the notion of experience is socialized (and socializing) through and through, reaffirming the experiencing subject's position on a historically contingent network of social relations, entities, and processes. Lawrence Grossberg (1992) introduces the notion of "mattering maps" to explain the way raw experience is socialized, pointing to the mechanisms by which affect, desire, and pleasure are invested in ideologically derived causes:

> The affective plain is organized according to maps which direct people's investments in and into the world. These maps are deployed in relation to the formations in which they are articulated. They tell people where, how and with what intensities they can become absorbed – into the world and

---

[35] See, for instance, Merleau-Ponty's (1962, p. 68) comment that in reality the phenomenal field and the transcendental field exist in mutuality.

their lives. This 'absorption' constructs the places and events which are, or can become, significant. They are the places at which people can anchor themselves into the world, the locations of the things that matter. (p. 82)

Mattering maps perform ideological transduction: they mobilize popular culture as a terrain upon which particular experiences are intentionally attributed with meaning. The same experience could be inflected by different, sometimes competing, mattering maps: one mattering map may chart a path toward pluralism or empathy, while another may channel emotional investment into violence, bigotry, or hate. Given that the cultural structures through which we articulate and make sense of our experiences are ideological, resonant interactions are equally amenable to competing ideological claims.[36] Seen in this light, sustainability itself appears as a kind of mattering map, without which the emotional investment generated by resonant interactions remains locked within the diegetic world. There is a case to be made, then, for coupling diegetic agency to extra-diegetic agency, even if the latter is merely suggestive or symbolic. We can imagine, for instance, a different oil spill installation where users may act (diegetically) to contain the spill, or a public engagement process in which resonant interactions are contextualized by different sustainability policymaking options and where users may enact real (extra-diegetic) political influence. In both cases, resonant interactions outline a different path for public engagement on sustainability. "Science can win minds, but you really need art to win hearts" (Rowsome 2017, May 10). How interactive media can assume an even more imaginative and imagination-inspiring role is the subject of the next chapter.

BIBLIOGRAPHY

Aarseth, E. J. (1997). *Cybertext: Perspectives on Ergodic Literature*. Baltimore: Johns Hopkins University Press.

Agamben, G. (1993). *Infancy and History: The Destruction of Experience* (trans: Heron, L.). London/New York: Verso.

Albert, B. (2016, November 14). Eagle Flight Review. *IGN*. Retrieved from http://www.ign.com/articles/2016/11/14/eagle-flight-review

---

[36] This is well illustrated by Jay's (2005, ch.5) discussion of the way experience was used in 1950–1960s Britain by both conservatives and Marxist Humanists to legitimate and consolidate diametrically opposing political positions.

Als, H. (2017, February 13 & 20). Capturing James Baldwin's Legacy Onscreen. *The New Yorker*. Retrieved from http://www.newyorker.com/ magazine/2017/02/13/capturing-james-baldwins-legacy-onscreen

Arthos, J. (2000). 'To Be Alive When Something Happens': Retrieving Dilthey's Erlebnis. *Janus Head*, 3(1). Retrieved from http://www.janushead.org/3-1/ jarthos.cfm

Ball, D. P. (2014, November 12). Burrard Inlet Binoculars Imagine Oil-Slicked Disaster. *The Tyee*. Retrieved from https://thetyee.ca/News/2014/11/12/ Burrard-Inlet-Installation/

Bendor, R. (2017). Interactive World Disclosure (or, an Interface Is Not a Hammer). In T. Markham & S. Rodgers (Eds.), *Conditions of Mediation: Phenomenological Perspectives on Media* (pp. 211–221). New York: Peter Lang.

Benjamin, W. (1929/1999). The Great Art of Making Things Seem Closer Together. In M. W. Jennings, H. Eiland, & G. Smith (Eds.), *Selected Writings (Vol. 2 Pt. 1)* (p. 248). Cambridge, MA: Belknap Press.

Bergson, H. (2007). *The Creative Mind: An Introduction to Metaphysics* (trans: Andison, M. L.). Mineola: Dover Publications.

Bord, R. J., O'Connor, R. E., & Fisher, A. (2000). In What Sense Does the Public Need to Understand Global Climate Change? *Public Understanding of Science, 9*, 205–218.

Broomell, S. B., Budescu, D. V., & Por, H.-H. (2015). Personal Experience with Climate Change Predicts Intentions to Act. *Global Environmental Change, 32*, 67–73.

Burdea, G., & Coiffet, P. (2003). *Virtual Reality Technology* (2nd ed.). Hoboken: J. Wiley-Interscience.

Candy, S. (2010). *The Futures of Everyday Life: Politics and the Design of Experiential Scenarios*. PhD dissertation submitted at the University of Hawaii, Manoa.

Chapman, D. A., Lickel, B., & Markowitz, E. M. (2017). Reassessing Emotion in Climate Change Communication. *Nature Climate Change, 7*, 850–852.

Chawla, L. (1998). Significant Life Experiences Revisited: A Review of Research on Sources of Environmental Sensitivity. *Environmental Education Research, 4*(4), 369–382.

Chawla, L. (2006). Learning to Love the Natural World Enough to Protect It. *Barn, 2*, 57–78.

Constine, J. (2017, January 20). Sundance Merges VR with Real Life Through Props, AR, and Vibrating Suits. *TechCrunch*. Retrieved from https:// techcrunch.com/2017/01/20/sundance-new-frontier/

Cox, J. R., & Pezzullo, P. C. (2015). *Environmental Communication and the Public Sphere* (4th ed.). Thousand Oaks: Sage.

Cresswell, T. (2015). *Place: An Introduction* (2nd ed.). Chichester/Malden: Wiley Blackwell.

Debord, G. (1994). *The Society of the Spectacle* (trans: Nicholson-Smith, D.). New York: Zone Books.

Deleuze, G., & Guattari, F. (1994). *What Is Philosophy?* (trans: Tomlinson, H., & Burchell, G.). New York: Columbia University Press.

Descartes, R. (2017). *Meditations on First Philosophy with Selections from the Objections and Replies* (trans: Cottingham, J., 2nd ed.). Cambridge: Cambridge University Press.

Dewey, J. (1934). *Art as Experience.* New York: Paragon.

Douglas, M., & Wildavsky, A. B. (1982). *Risk and Culture: An Essay on the Selection of Technical and Environmental Dangers.* Berkeley: University of California Press.

Dourish, P. (2001). *Where the Action Is: The Foundations of Embodied Interaction.* Cambridge, MA: MIT Press.

Egan, P. J., & Mullin, M. (2012). Turning Personal Experience into Political Attitudes: The Effect of Local Weather on Americans' Perceptions About Global Warming. *The Journal of Politics, 74*(3), 796–809.

Egan, P. J., & Mullin, M. (2017). Climate Change: US Public Opinion. *Annual Review of Political Science, 20*, 209–227.

Feenberg, A. (1999). *Questioning Technology.* London/New York: Routledge.

Feenberg, A. (2010). Between Reason and Experience. In *Between Reason and Experience: Essays in Technology and Modernity* (pp. 181–218). Cambridge, MA: MIT.

Fletcher, R. (2017). Gaming Conservation: Nature 2.0 Confronts Nature-Deficit Disorder. *Geoforum, 79*, 153–162.

Gordon, E., Schirra, S., & Hollander, J. (2011). Immersive Planning: A Conceptual Model for Designing Public Participation with New Technologies. *Environment and Planning B: Planning and Design, 38*, 505–519.

Grossberg, L. (1992). *We Gotta Get Out of This Place: Popular Conservatism and Postmodern Culture.* New York: Routledge.

Harman, G. (2007). *Heidegger Explained: From Phenomenon to Thing.* Chicago: Open Court.

Hart, P. S., & Leiserowitz, A. A. (2009). Finding the Teachable Moment: An Analysis of Information-Seeking Behavior on Global Warming Related Websites During the Release of the Day After Tomorrow. *Environmental Communication: A Journal of Nature and Culture, 3*(3), 355–366.

Heidegger, M. (1962). *Being and Time* (trans: Macquarrie, J., & Robinson, E.). San Francisco: HarperSanFrancisco.

Heidegger, M. (1971a). Building, Dwelling, Thinking. In *Poetry, Language, Thought* (trans: Hofstadter, A., pp. 145–161). New York: Harper and Row.

Heidegger, M. (1971b). The Thing. In *Poetry, Language, Thought* (trans: Hofstadter, A., pp. 165–182). New York: Harper and Row.

Hornsey, M. J., Harris, E. A., Bain, P. G., & Fielding, K. S. (2016). Meta-Analyses of the Determinants and Outcomes of Belief in Climate Change. *Nature Climate Change, 6*, 622.

Ihde, D. (1990). *Technology and the Lifeworld: From Garden to Earth*. Bloomington: Indiana University Press.

Ingold, T. (2000). *The Perception of the Environment: Essays on Livelihood, Dwelling and Skill*. London/New York: Routledge.

IPCC. (2014). *Climate Change 2014 Mitigation of Climate Change: Working Group III Contribution to the Fifth Assessment Report of the Intergovernmental Panel on Climate Change*. New York: Cambridge University Press.

Jaffe, Z. (2017, February 23). Tree VR "Grows" with SUBPAC at Sundance. Retrieved from http://subpac.com/tree-vr-grows-subpac-sundance/

Jang, S. M., & Hart, P. S. (2015). Polarized Frames on "Climate Change" and "Global Warming" Across Countries and States: Evidence from Twitter Big Data. *Global Environmental Change, 32*, 11–17.

Jay, M. (2005). *Songs of Experience: Modern American and European Variations on a Universal Theme*. Berkeley: University of California Press.

Jégou, F., & Gouache, C. (2015). Envisioning as an Enabling Tool for Social Empowerment and Sustainable Democracy. In V. W. Thoresen, R. J. Didham, J. Klein, & D. Doyle (Eds.), *Responsible Living: Concepts, Education and Future Perspectives* (pp. 253–271). Cham: Springer.

Joshi, K. (2017, June 7). Caring About Climate Change: It's Time to Build a Bridge Between Data and Emotion. *The Guardian*. Retrieved from https://www.theguardian.com/commentisfree/2017/jun/07/caring-about-climate-change-its-time-to-build-a-bridge-between-data-and-emotion

Kitchin, R., & Freundschuh, S. (2000). *Cognitive Mapping: Past, Present and Future*. London: Routledge.

Lakoff, G. (2004). *Don't Think of an Elephant!: Know Your Values and Frame the Debate: The Essential Guide for Progressives*. White River Junction: Chelsea Green Pub. Co.

Lakoff, G., & Johnson, M. (1980/2003). *Metaphors We Live by*. Chicago: University of Chicago Press.

Lawson, P. J., & Flocke, S. A. (2009). Teachable Moments for Health Behavior Change: A Concept Analysis. *Patient Education and Counseling, 76*, 25–30.

Lessig, L. (2006). *Code: Version 2.0* (2nd ed.). New York: Basic Books.

Li, Y., Johnson, E. J., & Zaval, L. (2011). Local Warming: Daily Temperature Change Influences Belief in Global Warming. *Psychological Science, 22*(4), 454–459.

MacKinnon, J. B. (2013). *The Once and Future World: Nature as It Was, as It Is, as It Could Be*. Boston/New York: Houghton Mifflin Harcourt.

Malpas, J. (2006). *Heidegger's Topology: Being, Place, World*. Cambridge, MA: MIT Press.

Mason, S. (2013). On Games and Links: Extending the Vocabulary of Agency and Immersion in Interactive Narratives. In H. Koenitz, T. I. Sezen, G. Ferri, M. Haahr, D. Sezen, & G. v. Çatak (Eds.), *Interactive Storytelling* (pp. 25–34). Heidelberg: Springer.

Maxwell, R. (2013, March 8). Spatial Orientation and the Brain: The Effects of Map Reading and Navigation. *GIS Lounge*. Retrieved from https://www.gislounge.com/spatial-orientation-and-the-brain-the-effects-of-map-reading-and-navigation/

McCarthy, J., & Wright, P. (2004). *Technology as Experience*. Cambridge, MA: MIT Press.

McMahan, A. (2003). Immersion, Engagement, and Presence: A Method for Analyzing 3-D Video Games. In M. J. P. Wolf & B. Perron (Eds.), *The Video Game Theory Reader* (pp. 67–86). New York/London: Routledge.

Merleau-Ponty, M. (1962). *Phenomenology of Perception* (trans: Smith, C.). London/Henley: Routledge/Kegan Paul.

Minsky, M. (2006). *The Emotion Machine: Commonsense Thinking, Artificial Intelligence, and the Future of the Human Mind*. New York: Simon & Schuster.

Moores, S. (2012). *Media, Place and Mobility*. Basingstoke/New York: Palgrave Macmillan.

Moser, S. C., & Dilling, L. (2007). Introduction. In S. C. Moser & L. Dilling (Eds.), *Creating a Climate for Change: Communicating Climate Change and Facilitating Social Change* (pp. 1–27). Cambridge/New York: Cambridge University Press.

Moser, S. C., Daniels, C., Pike, C., & Huva, A. (2016). *Here-Now-Us: Visualizing Sea Level Rise and Adaptation Using the OWL Technology in Marin County*, Santa Cruz. Retrieved from https://climateaccess.org/sites/default/files/Here%20Now%20Us%20Project%20and%20Research%20Summary.pdf

Murray, J. H. (1997). *Hamlet on the Holodeck: The Future of Narrative in Cyberspace*. Cambridge, MA: MIT Press.

Myers, T. A., Maibach, E. W., Roser-Renouf, C., Akerlof, K., & Leiserowitz, A. A. (2013). The Relationship Between Personal Experience and Belief in the Reality of Global Warming. *Nature Climate Change, 3*, 343–347.

O'Neil, P. (2014, April 30). Kinder Morgan Pipeline Application Says Oil Spills Can Have Both Negative and Positive Effects. *Vancouver Sun*. Retrieved from http://www.vancouversun.com/news/Kinder+Morgan+pipeline+application+says+spills+have+both+negative+positive+effects/9793673/story.html

Otto, E. C., & Wilkinson, A. (2012). Harnessing Time Travel Narratives for Environmental Sustainability Education. In A. E. J. Wals & P. B. Corcoran (Eds.), *Learning for Sustainability in Times of Accelerating Change* (pp. 91–104). Wageningen: Wageningen Academic Publishers.

Patterson, R., Winterbottom, M. D., & Pierce, B. J. (2006). Perceptual Issues in the Use of Head-Mounted Visual Displays. *Human Factors: The Journal of the Human Factors and Ergonomics Society, 48*(3), 555–573.

Regan, T. (2016, November 22). Eagle Flight. *Trusted Reviews*. Retrieved from http://www.trustedreviews.com/reviews/eagle-flight

Reibelt, L. M., Richter, T., Rendigs, A., & Mantilla-Contreras, J. (2017). Malagasy Conservationists and Environmental Educators: Life Paths into Conservation. *Sustainability, 9*(2), article #227.

Relph, E. (2008). *Place and Placelessness* (Reprinted ed.). London: Pion Limited.

Renn, O. (2011). The Social Amplification/Attenuation of Risk Framework: Application to Climate Change. *Wiley Interdisciplinary Reviews: Climate Change, 2*(2), 154–169.

Reser, J. P., Morrissey, S. A., & Ellul, M. (2011). The Threat of Climate Change: Psychological Response, Adaptation, and Impacts. In I. Weissbecker (Ed.), *Climate Change and Human Well-Being: Global Challenges and Opportunities* (pp. 19–42). New York/Dordrecht/Heidelberg/London: Springer.

Reser, J. P., Bradley, G. L., & Ellul, M. C. (2014). Encountering Climate Change: 'Seeing' Is More Than 'Believing'. *Wiley Interdisciplinary Reviews: Climate Change, 5*(4), 521–537.

Rheingold, H. (1991). *Virtual Reality.* New York: Summit Books.

Robertson, A. (2017, January 26). The Best Virtual Reality from the 2017 Sundance Film Festival. *The Verge.* Retrieved from https://www.theverge.com/2017/1/26/14396976/best-vr-sundance-film-festival-2017

Rorty, R. (2007). Philosophy as a Transitional Genre. In *Philosophy as Cultural Politics* (pp. 3–28). Cambridge: Cambridge University Press.

Rowsome, A. (2017, May 10). Can Virtual Reality Help Us Tackle Climate Change? *Vice Impact.* Retrieved from https://impact.vice.com/en_us/article/xyeg97/can-virtual-reality-help-us-tackle-climate-change

Sapieha, C. (2016, November 9). Eagle Flight Review: Time Keeps on Slipping in Ubisoft's Repetitive VR Simulation of Bird Life. *Financial Post.* Retrieved from http://business.financialpost.com/technology/gaming/eagle-flight-review-time-keeps-on-slipping-in-ubisofts-repetitive-vr-simulation-of-bird-life

Sassoon, D. (2012, August 20). Crude, Dirty and Dangerous. *The New York Times.* Retrieved from http://www.nytimes.com/2012/08/21/opinion/the-dangers-of-diluted-bitumen-oil.html

Schroeder, R. (1996). *Possible Worlds: The Social Dynamic of Virtual Reality Technology.* Bolder: Westview Press.

Schueneman, T. (2013, July 26). Future of San Francisco's Market Street Comes into View. *Triple Pundit.* Retrieved from http://www.triplepundit.com/2013/07/future-san-franciscos-market-street-comes-view/

Schuldt, J. P., & Roh, S. (2014). Of Accessibility and Applicability: How Heat-Related Cues Affect Belief in "Global Warming" Versus "Climate Change". *Social Cognition, 32*(3), 217–238.

Schuldt, J. P., Konrath, S. H., & Schwarz, N. (2011). "Global Warming" or "Climate Change"?: Whether the Planet Is Warming Depends on Question Wording. *Public Opinion Quarterly, 75*(1), 115–124.

Scott, J. C. (1998). *Seeing Like a State: How Certain Schemes to Improve the Human Condition Have Failed.* New Haven/London: Yale University Press.

Seamon, D. (1979). *A Geography of the Lifeworld: Movement, Rest and Encounter.* London: Croom Helm.

Seamon, D. (2015). Situated Cognition and the Phenomenology of Place: Lifeworld, Environmental Embodiment, and Immersion-in-World. *Cognitive Processing, 16*(supplement 1), 389–392.

Sheppard, S. R. J. (2005). Landscape Visualisation and Climate Change: The Potential for Influencing Perceptions and Behaviour. *Environmental Science and Policy, 8,* 637–654.

Stevenson, K. T., Peterson, M. N., Carrier, S. J., Strnad, R. L., Bondell, H. D., Kirby-Hathaway, T., & Moore, S. E. (2014). Role of Significant Life Experiences in Building Environmental Knowledge and Behavior Among Middle School Students. *The Journal of Environmental Education, 45*(3), 163–177.

Stinson, J. (2017). Re-creating Wilderness 2.0: Or Getting Back to Work in a Virtual Nature. *Geoforum, 79,* 174–187.

Tanenbaum, J. (2014). Design Fictional Interactions: Why HCI Should Care About Stories. *Interactions, 21*(5), 22–23.

Thiel, P. (1996). *People, Paths, and Purposes: Notations for a Participatory Envirotecture.* Seattle/London: University of Washington Press.

Tomlinson, B. (2010). *Greening Through IT: Information Technology for Environmental Sustainability.* Cambridge, MA: MIT Press.

Tuan, Y.-f. (1977). *Space and Place: The Perspective of Experience.* Minneapolis: University of Minnesota Press.

Vaillant, J. (2005). *The Golden Spruce: A True Story of Myth, Madness and Greed.* New York: W.W. Norton.

Wallace-Wells, D. (2017, July 9). The Uninhabitable Earth. *New York Magazine.* Retrieved from http://nymag.com/daily/intelligencer/2017/07/climate-change-earth-too-hot-for-humans.html

Weber, E. U. (2006). Experience-Based and Description-Based Perceptions of Long-Term Risk: Why Global Warming Does Not Scare Us (Yet). *Climatic Change, 77*(1–2), 103–120.

Weber, E. U. (2010). What Shapes Perceptions of Climate Change? *Wiley Interdisciplinary Reviews: Climate Change, 1*(3), 332–342.

Weber, E. U. (2016). What Shapes Perceptions of Climate Change? New Research Since 2010. *Wiley Interdisciplinary Reviews: Climate Change, 7*(1), 125–134. https://doi.org/10.1002/wcc.377.

Weisman, A. (2007). *The World Without Us.* New York: Picador.

Wells, N. M., & Lekies, K. S. (2006). Nature and the Life Course: Pathways from Childhood Nature Experiences to Adult Environmentalism. *Children, Youth and Environments, 16*(1), 1–24.

Whitmarsh, L. (2008). Are Flood Victims More Concerned About Climate Change Than Other People? The Role of Direct Experience in Risk Perception and Behavioural Response. *Journal of Risk Research, 11*(3), 351–374.

Whitmarsh, L. (2009). What's in a Name? Commonalities and Differences in Public Understanding of "Climate Change" and "Global Warming". *Public Understanding of Science, 18*(4), 401–420.

Williams, C. C., & Chawla, L. (2016). Environmental Identity Formation in Nonformal Environmental Education Programs. *Environmental Education Research, 22*(7), 978–1001.

Wynne, B. (1992). Misunderstood Misunderstanding: Social Identities and Public Uptake of Science. *Public Understanding of Science, 1*(3), 281–304.

CHAPTER 5

Imagination

POLITICAL AND ECONOMIC MYOPIA

A survey of residents of British Columbia (BC), Canada, done by Justason Market Intelligence Inc. in January 2014, found that despite the fact that 64% of respondents opposed allowing crude oil supertankers to pass through BC's inside passage, part of Enbridge Corporation's Northern Gateway plan, a similar percentage of respondents believed that the plan will likely be allowed to go ahead anyway.[1] An even higher percentage of respondents (79%) told pollsters that the public should actively participate in decision-making over the project instead of leaving the government to make the decision on its own. The results indicated that the project lacked social license, a significant impediment given that roughly half of the thousand-kilometer pipeline was to stretch across BC territory (and pass through the unceded territories of several First Nations), and that the project also included constructing a large shipping terminal at the northern BC town of Kitimat. Apparently, the project's estimated job gains and tax revenue for BC did not seem worth the risks involved in moving crude

---

[1] For more on the project, see Enbridge's website (https://www.enbridge.com). The poll was commissioned by the Dogwood Initiative, ForestEthics Advocacy, the Northwest Institute for Bioregional Research, and West Coast Environmental Law. A report on the survey can be found here: http://www.justasonmi.com/wp-content/uploads/2014/02/January-2014-Pipelines-and-Tanker-Proposal-charts.pdf (last accessed Mar. 18, 2018).

© The Author(s) 2018
R. Bendor, *Interactive Media for Sustainability*,
Palgrave Studies in Media and Environmental Communication,
https://doi.org/10.1007/978-3-319-70383-1_5

oil across the Canadian Rockies, and in hosting annually more than 200 oil tankers that would have to navigate BC's treacherous waters before and after arriving at Kitimat.

Based on the fact that 51% of respondents claimed to "distrust" the Canadian federal government's approval process, the survey reflects a deep disillusionment with the federal government's policies—a curious fact given the ideological alignment between the province's Liberal government and the federal Conservatives who were in power at the time.[2] Doubtlessly, this mistrust could be traced to the persistence of belief in BC that while the interests of the federal government extend all the way to the Pacific Coast, fair treatment of provinces ends at Ontario's western border. But the survey may indicate something else. Given the public's desire to influence the process and almost equally strong belief that they are unable to actually do so, the survey reveals a worrying lack of perceived political self-efficacy. Whether this reflects public skepticism about the political system's amenability to change, or a lack of belief in the effectiveness of public means to precipitate political change (Caprara et al. 2009), it seems clear that those surveyed felt powerless in the face of fossil-fueled politics, incapable of imagining and pursuing impactful political processes that would reign in the latest incarnation of "fossil capital" (Malm 2016).[3] It is not only that, as one anti-pipeline campaigner remarked, "our democratic system is broken" (cited in Klein 2014, p. 363), but that we are finding it hard to imagine how it can be fixed. Legal challenges and the declining price of oil have since rendered Northern Gateway all but dead,[4] but given that other plans to build or expand pipelines in the province (and elsewhere) are still very much alive, questioning the political imagination is still both relevant and urgent. Are we indeed in the thralls of what some have called a "crisis of imagination" (Ghosh 2016, Oct. 28; Haiven 2014; Monbiot 2017, Sep. 9; Wals and Corcoran 2012)? And if so, what can be done about it?

---

[2] Trust in the provincial government was much more robust: nearly half of those surveyed said they would be more supportive of the project if the province approved it first.

[3] Of course not all BC residents feel politically impotent. Several civil society groups have been and still are very active in opposing similar projects, such the Trans Mountain Pipeline, with varying degrees of success.

[4] 'Canada's Crude Oil Pacific Exports and Northern Gateway Likely Dead if Proposed Tanker Ban Becomes Law': https://www.oilandgas360.com/canadas-crude-oil-pacific-exports-northern-gateway-likely-dead-proposed-tanker-ban-becomes-law (last accessed Mar. 18, 2018).

Similar questions emerge from a different survey. At the end of 2011, and in anticipation of the Rio+20 Conference on Sustainable Development, 650 sustainability business experts and practitioners from around the world were asked about what they perceived to be the most important barriers preventing the world's progress on sustainability.[5] The most popular answer, and by a significant margin, was "financial short-termism." The UN's Global Compact ("The world's largest corporate sustainability initiative") describes the problem as follows:

> The short-term performance pressures on investors result in an excessive focus on quarterly earnings, with less attention paid to strategy, fundamentals and long-term value creation. Many companies respond to these pressures by reducing expenditures on research and development and foregoing investment opportunities with a positive long-term net present value. As a result, companies are discouraged from developing sustainable products, investing in measures that deliver operational efficiencies, developing their human capital, or effectively managing the social and environmental risks to their business.[6]

Simply put, businesses are letting short-term goals dictate their long-term strategies, resulting in negative impact on their overall value, their environmental performance, and society at large. Various reports published after the 2011 survey confirm that the tendency to sacrifice the future at the altar of quarterly gains is still very much alive.[7]

One may be inclined to pin the financial discounting of the future on the practices of particularly susceptible financial institutions, "where investment managers are hired, compensated and fired based on their performance over 6–12 months" (Erikson 2012, Mar. 5). However, given the prevalence of short-termism across the corporate world, it can equally be seen as a structural failure, something that is deeply embedded in the very way businesses approach their mandate and

---

[5] The survey is reported in Erikson (2012, Mar. 5). It was conducted by GlobeScan and SustainAbility, in collaboration with the United Nations Environment Programme.

[6] 'Short-Termism in Financial Markets': https://www.unglobalcompact.org/take-action/ action/long-term (last accessed Mar. 18, 2018).

[7] For recent examples, see 'LEAD Companies tackle the dilemma of short-termism': https://www.unglobalcompact.org/news/791-01-22-2014 (last accessed Mar. 18, 2018); and 'Closing the Sustainability-Investor Relations Gap': http://10458-presscdn-0-33. pagely.netdna-cdn.com/wp-content/uploads/2016/12/SA-ES-Report-Web-Spreads.pdf (last accessed Mar. 18, 2018).

responsibilities—what Charles Eisenstein (2011) calls "the myopia of capital." Seen this way, what we are witnessing is much more than the effects of inappropriate financial regulations (listed as the second most important barrier to progress on sustainability in the aforementioned survey), not a failure of corporate practices but one of the corporate imagination.

Taking into account both surveys, what emerges is a real sense that the path to sustainability is obstructed by our own inability as individuals and as a collective to imagine what a sustainable future may look like. We are facing a crisis of the imagination, or more accurately, crises of our social, economic, and political imaginaries (Taylor 2004).[8] Identifying such crises reflects the degree to which sustainability relies on deeper cultural structures, values, and beliefs (Ehgartner et al. 2017; Ehrenfeld 2008; Lövbrand et al. 2015; Maggs and Robinson 2016). From this perspective, what sustainability means in both theory and practice is premised in how we make sense of the world and ourselves. But no less important, recognizing these crises *as such* prompts an urgent need to develop ways to engage, prop up, evoke, or otherwise unleash the public's imagination (Bendor 2018; Ghosh 2016, Oct. 28; McNichol 2010; Wals and Corcoran 2012; Wright et al. 2013; Yusoff and Gabrys 2011). As George Monbiot (2017, Sep. 9) writes, "As we rekindle our imagination, we discover our power to act." The interactive media discussed in this chapter address this challenge by deploying what I have called elsewhere *worldmaking interactions*: forms of interaction that "aim to promote the public's own ability to imagine alternative futures – to encourage the public to find ways to collectively reformulate a sense of what is possible and hopefully rediscover its capacity to critically make, unmake and remake the world" (Bendor 2018, p. 206).[9] Worldmaking interactions aim to evoke and create traffic between the individual's imagination and the more collective, social imaginaries. They are unabashedly experimental, even speculative in nature, yet seem primed to unlock one of the more important aspects of politics. How they do so in the context of sustainability is the subject of this chapter.

---

[8] Unless noted otherwise, I will refer to all these sub-types as a singular "social imaginary," or simply, "imaginary." But this should not obscure the fact that there are indeed different types of imaginaries and different imaginaries of the same type.

[9] I borrow the term "worldmaking" from Nelson Goodman (1978). For a lengthier discussion of worldmaking, see Vervoort et al. (2015).

Unsettling the World

Both surveys discussed above project a strong sense of inevitability. In the first, although the public clearly reject the proposed pipeline, they feel that the project's momentum is unstoppable. In the second, despite recognizing the counter-productivity of financial short-termism, business cannot find levers for change. In both cases, then, the pull of the "here and now" seems too strong to overcome, or, in other words, the dominance of existing social imaginaries seems to leave no room for alternatives. No stars are visible in the night skies of fossil capital.

Social imaginaries, explains Charles Taylor (2004), are "the ways people imagine their social existence, how they fit together with others, how things go on between them and their fellows, the expectations that are normally met, and the deeper normative notions and images that underlie these expectations" (p. 23). Haiven and Khasnabish (2014, p. 4) add that social imaginaries are "multiple, overlapping, contradictory and coexistent imaginary landscapes, horizons of common possibility and shared understanding." They provide the basis for moral judgment and juridical legitimacy, outline desirable forms of social organization and governance, and orient everyday life. Manifesting something like "culture's choreography of contexts" (as Tresch (2007, p. 93) writes about "cosmograms"), social imaginaries mix concepts and practice, historical narratives and future aspirations, to provide a sense of the world and the relations it holds. They are formed through the accretion of beliefs, ideas, and everyday activities, the stories we tell ourselves about who we are, where we came from, where we are going, and why. In this sense, social imaginaries provide society with a common vision of itself and thus help to integrate it further. As Cornelius Castoriadis (1997) argues, the social imaginary is not merely "something in the world" but the repository of potentiality from which something may emerge—"the unceasing and essentially *undetermined* (social-historical and psychical) creation of figures/forms/images, on the basis of which alone there can ever be a question *of* 'something'. What we call 'reality' and 'rationality' are its works" (p. 3; emphasis in origin). Continuously changing, social imaginaries simultaneously derive from and shape reality. They manifest our horizons of possibility.

Taylor (2004) points out that our social imaginaries often remain unarticulated because, in effect, they are inarticulable. They are unlimited and indefinite, available only as background structures against which "particular features of our world show up for us in the sense they have" (p. 25).

The perceived ephemerality of the social imaginary may be used to discount it as a "wooly feel-good slogan that distracts us from the 'real' work of social justice" (Haiven and Khasnabish 2014, p. 3); however, as Naomi Klein (2014) explains, the success of transformative social movements often depends on their ability to contest existing imaginaries and offer alternatives:

> all of them [successful social movements] understood that the process of shifting cultural values—though somewhat ephemeral and difficult to quantify—was central to their work. And so they dreamed in public, showed humanity a better version of itself, modeled different values in their own behavior, and in the process liberated the political imagination and rapidly altered the sense of what was possible. (p. 462)[10]

This substantiates Klein's argument that the overarching task of a global ecological movement "is to articulate not just an alternative set of policy proposals but an *alternative worldview to rival the one at the heart of the ecological crisis*—embedded in interdependence rather than hyper-individualism, reciprocity rather than dominance, and cooperation rather than hierarchy" (ibid.; emphasis added). If Klein is correct and struggles for social transformation are indeed waged first in the realm of the imaginary, this would explain, on the one hand, the efforts of hegemonic political actors to limit the reach and purchase of the "radical imagination" (Haiven and Khasnabish 2014) and, on the other hand, the current drive by progressives to theorize and mobilize social imaginaries that contrast and challenge those that serve to maintain late capitalism and its fossil politics-as-usual (see, e.g., Augé 2015; Bottici 2014; Ghosh 2016, Oct. 28; Williams and Srnicek 2013). If our social imaginaries are both the ground from which social change sprouts and the horizon of possibility that limits the reach and depth of social change, they are not only an object worthy of politics but the *objet politique par excellence*.

"Is This the Kind of Life You Imagine Living?"

Having identified the pivotal role played by the social imaginary in catalyzing political, economic, and cultural transformation, we may ask how we could affect it. This is complicated by the fact that the qualities of

---

[10] In Steve Lambert's words: "All effective activism has to be artistic – it needs to innovate, capture the imagination, envision new futures, and present the world in a way people can't otherwise see" (cited in Alvarez 2017, Apr. 26).

any particular imaginary may be difficult to discern, that society may feature more than one imaginary, and that the presence of alternative (rival, radical, subversive) imaginaries may remain hidden from those inhabiting the social space outlined by the dominant imaginary. Nonetheless, the ephemerality of the social imaginary does not mean that it could not be affected, rejected, or even transformed. History provides us with ample examples of that. Perhaps the most intuitive way by which social imaginaries may be affected is by providing the imagination with inspiring examples of alternatives. In Chris Turner's *The Geography of Hope: A Tour of the World We Need* (2007), for instance, we find multiple examples not only of what *can* be done to combat climate change, but also of what is *already* done, proof that different imaginaries are already operative in pursuit of transformative change. Similarly, as Cross et al. (2015) find, media stories about successful green entrepreneurial activism and "everyday heroism" provide more motivation for popular participation in climate politics than stories that aim to teach or preach about the sources and consequences of climate change. Such stories light up the imagination with concrete possibility.

Interactive media can help. Take, for example, the Swedish online interactive exhibition *Life 2053*, which provides visitors a glimpse of "sustainable life in a major Swedish city in 2053"[11]:

> The year is 2053. Energy use is 60% lower than in 2000. The last 50 years have seen major technological advances in energy efficiency, which along with behavioural changes have led to energy consumption decreasing from 33 MWh to 14 MWh per person and year in 2053.

Visitors can read fictional news from the future ("Peak wood warning!," "Meat smuggling scandal!"), and are asked to make several choices about their preferred lifestyle in the future (regarding weekly working hours, income, housing, etc.). Visitors can then see how their choices would play out in a series of visual vignettes about food preparation, commuting to work, clothing, and vacationing. Each vignette features fictional conversations with other characters, additional choice making, and examples of how new domestic technologies will affect the lifestyle of future urban dwellers. At the end of the presentation, visitors receive a

---

[11] The exhibition was created by Green Leap at KTH Royal Institute of Technology, and supported by the Swedish Energy Agency in cooperation with Centre for Sustainable Communications, KTH. It can be visited at http://www.life2053.com (last accessed Mar. 18, 2018).

breakdown of their future energy consumption based on in-game choices. This sets up the exhibition's key question: "Is this the kind of life you imagine living?"

The exhibition conveys a strong sense of achievability by projecting into the future existing measures to reduce the urban ecological footprint. Its anticipation of "Less flights and energy-efficient aircraft using biofuels," "Shorter commutes, with increased share of public transportation," "More energy efficient homes," and "Less meat and more seasonal food" can all be seen as upshots of the exhibition's adherence to what was referred to in the previous chapter as scientific defensibility. The price the exhibition pays for conveying achievability, however, is that it does not seem to go far enough to suggest real alternative imaginaries. The existence and availability of "smart" appliances, small urban food production facilities, means for teleconferencing, elevated bike lanes, or even tool and car sharing, is not entirely of the future. Much of it already exists. Furthermore, because the exhibition locates these largely technological means in what amounts to fairly contemporary social practices, it produces a sense of comfort with the future—a comfort that both indicates and promulgates a certain satisfaction with current social imaginaries (or at least with the future trajectories they harbor). The imagined future the exhibition presents is not so radically different from the present. On the one hand, this makes sustainability—the version of sustainability championed by the exhibition creators, that is—seem realistically achievable, firmly within our collective reach. On the other hand, it does so without significantly challenging our current ways of being. Humans, the exhibition tells us, will solve the climate crisis without making significant sacrifices: no need to challenge the values or cultural norms promoted by neoliberal capitalism; a more sustainable future is not only within our technological reach but is almost certainly on the way. In this sense, *Life 2053* does more to *extend* our imaginaries than to fundamentally challenge them.

CHALLENGING SCIENTIFIC IMAGINARIES

Life 2053, as we have seen, makes use of scientific data to ground its calculations and projections. It shrouds itself in scientific rhetoric to convey a sense of realistic possibility, but it does not challenge the kind of imaginaries it extends. Is this a missed opportunity? Perhaps, since, historically speaking, challenges to scientific imaginaries have been a major source of social transformation. Copernicus's heliocentrism, Newton's mechanics, Darwin's evolution, Pasteur's microbes, Einstein's relativity, and Freud's unconscious (if we are to take Freud's own immodest word for it)[12] provide a few examples of how existing imaginaries struggle and ultimately fail to survive the impact of scientific revolutions (Kuhn 1962).[13]

For some, the growing scientific cachet of the Anthropocene appears to offer a similarly unsettling "mind bomb" that would compel society not only to recognize and address the scale of the material outcomes of modern industrial capitalism but also to recalibrate what it means to be human. This would imply, at the very least, a significant challenge to scientific practices and epistemologies (Trischler 2016). But the conceptual pull of the Anthropocene may also inspire political reconfigurations (Bonneuil and Fressoz 2017; Latour 2017, pp. 143–144; Mann and Wainwright 2018), and even motivate a deep ontological shift. In the context of the latter, Bruno Latour (2014) believes that the notion would help to dispatch with the nature-culture rift altogether (the main thrust of his theoretical oeuvre).[14] He writes:

> The point of living in the epoch of the Anthropocene is that all agents share the same shape-changing destiny. A destiny that cannot be followed, documented, told, and represented by using any of the older traits associated with subjectivity or objectivity. Far from trying to 'reconcile' or 'combine' nature and society, the task, the crucial political task, is on the contrary to distribute agency as far and in as differentiated a way as possible – until, that is, we have thoroughly lost any relation between those two concepts of object and subject that are of no interest any more except patrimonial. (p. 17)

In less philosophical terms, the novelist Amitav Ghosh (2016, Oct. 28) argues that the Anthropocene presents a challenge "to our common sense

---

[12] See Freud (1920, pp. 246–247).

[13] See, for instance, Dijksterhuis's (1961) account of the dramatic changes introduced at the period that stretches between Copernicus and Newton, on which he writes: "It brought about an enormous advance in men's knowledge and technical skill, and consequently a radical change in their views of life and of the world" (p. 287). Of course, some beliefs may survive even the most fundamental scientific revolutions, as evident in the persistence of "Flat Earth" societies.

[14] See in particular *We Have Never Been Modern* (1993), *Politics of Nature* (2004), and *Facing Gaia* (2017).

understandings and beyond that to contemporary culture in general."
Garrard et al. (2014) similarly suggest that:

> The concept of the Anthropocene asks that we think and imagine on a
> wholly different scale, vastly more global in scope, vastly more historical in
> extent, in the course of making decisions about countless matters of envi-
> ronmental concern. And it asks that we take seriously the specific responsi-
> bilities that arise from this shifting of perspectives. (p. 150)

We may say that the view of humanity as a geological force—Homo Faber
as a force of nature—demands that we examine and check our civilizational hubris. It presents us with an opportunity to tell ourselves different
stories about who we are, what we can, and what we should do. It is, in
other words, an opportunity for both an ontological re-evaluation and an
ethical readjustment, both of which carry important implications for sustainability (Arias-Maldonado 2013; Maggs and Robinson 2016).

Whether or not the Anthropocene will indeed spark the kind of ontological transformation Latour anticipates, it provides an opportunity for
societal soul searching—a measure of reckoning with humanity's capacity
to inflict changes to the very fabric of the planet, and the planet's capacity
to retaliate. One example of how such a reckoning may be provoked is
Tamiko Thiel's interactive installation, *Gardens of the Anthropocene*. The
installation was originally exhibited during the summer of 2016 in the
Seattle Art Museum's Olympic Sculpture Park, but has since "migrated"
and could be experienced in other locations including Stanford, California,
and Brooklyn, New York. The installation is entirely virtual and could only
be experienced by using the augmented reality (AR) application Layar on
a mobile device. Once users install the app, and when they are located in
the appropriate site, they can experience different types of mutated plants
superimposed dynamically on the surrounding ("real") environment (see
Image 5.1).[15]

The installation was described by the Museum as follows:

> In Thiel's virtual world, native plants grow and mutate in response to the
> earth's changing conditions, adapting to the evolving climate and altering
> the landscape as we know it. Some of the transformations envisioned by the

---

[15] Short videos depicting the installation as viewed through the Layar application can be
found here: https://vimeo.com/177393844 and here: https://vimeo.com/190628682
(last accessed Mar. 18, 2018).

Image 5.1   Gardens of the Anthropocene, augmented reality installation by Tamiko Thiel. (Image by Tamiko Thiel)

artist include bullwhip kelp flying high above sea level or street signs whirling in the air. While these images may seem surreal, they point to a dystopian future where the natural world has been irrevocably impacted by mankind.[16]

Thiel chose to work with plants that were both native to the Museum's location and scientifically recognized for their high adaptability. Plant mutations were based on real botanical information, projected into a future underlined by climate change and erratic weather events. Each plant was given a Latin name that manifests its future mutation, so the speculative mutation of bullwhip kelp (*Nereocystis luetkeana*), for instance, was now called *Nereocystis volans* to reflect the way it could fly in the air like a drone. Alongside new airborne kelp species, we can find other mutated plant varieties including *Clarkia antenna* (a pinkish flower that

---

[16] 'Tamiko Thiel: Gardens of the Anthropocene': http://seattleartmuseum.org/gardens (last accessed Mar. 18, 2018).

feeds on electromagnetic signals and has antennae instead of pollinators); *Camassia radaria* (a blue flower that upon becoming agitated by the presence of humans spins like a radar antenna); *Alexandrium giganteus, aerius,* and *colossus* (three varieties of supersized land-living "red tide" algae); and *Pseudo-nitzschia immensa* (a form of plankton that causes short-term memory loss to the fish that consume it).

The new mutations not only represent floral responses to a changing climate but also challenge a variety of biological distinctions that seem to buckle under the symbolic weight of the Anthropocene. Thiel makes this point on her website:

> the artwork takes artistic license to imagine a surreal, dystopian scenario in which plants are "mutating" to breach natural boundaries: from photosynthesis of visible light to feeding off of mobile devices' electromagnetic radiation, from extracting nutrients from soil to feeding off man-made structures, and to transgressing boundaries between underwater and dry land, between reactive flora and active fauna.[17]

Transgressing these boundaries echoes the way the Anthropocene itself blurs nature-culture distinctions. If humans have become a force of nature, it only stands to reason that nature will adapt and take advantage of human presence (of course, in many ways it already has). The result, however, is neither strictly "natural" nor "cultural." In terms of its content, then, the installation mobilizes nonorthogenetic botany to formulate and materialize a new hybrid biological imaginary fit for the Anthropocene. It draws from scientific imaginaries while transcending them, performing an artistic jiu-jitsu. The installation's form complements the installation's content. Compared to the media discussed in the previous chapter, *Gardens* seems rudimentary, almost a caricature. While interactive installations such as *Tree,* and VR games such as *Eagle Flight* pursue high-fidelity aesthetics as a way to achieve immersive realism, *Gardens* seems satisfied with its artificiality. In this sense, the installation's speculative scientific premise feeds into its aesthetics; its grotesque depiction of techno-natural objects and refusal to hide the nonvirtual environment underneath the virtual images reveal the installation itself as artifice—more a case of magical than scientific realism. *Gardens'* aesthetic, in other words, is essentially unfinished, functioning more as a scaffold for the imagination than a complete world to inhabit. I will return to the question of unfinishedness in more detail following a brief discussion of the general characterisics of the imagination.

---

[17] http://tamikothiel.com/gota (last accessed Mar. 18, 2018).

OUTLINES OF THE IMAGINATION

The imagination is a curious thing. On the one hand, it is lauded as the seat of creativity, that which allows us to "transcend the *actual* and project ourselves into the *possible*" (Kearney 1998, p. 76; emphasis in origin). As epitomized in Baudelaire's words: "Imagination created the world" (cited in ibid., p. 5). On the other hand, the imagination is vilified as an impoverished mode of perceiving the world, a source of fictitious, fanciful, and erroneous knowledge. Plato located the imagination at the bottom of his hierarchy of mental faculties, the Latin author Pliny decreed that "nothing could be more foolish than a man ruled by imagination" (cited in ibid., p. 3), and Husserl asserted that the imagination "can 'presentify' ... yet cannot achieve the amplitude and plentitude that characterize sensory perception" (Casey 1976, p. 2). Rich but impoverished; creative but fanciful. Perhaps Kant was right when he wrote that the imagination is both "blind" but "indispensable" (ibid., p. 3). In any case, the aim of this section is not to solve or untangle this "duality of image-making" (Kearney 1998, p. 3) but to illustrate some of its constituents. The duality of the imagination will not be treated here as a deficiency but rather as an opportunity.

For individuals, the imagination is a means to conjure, realize, and bring into existence. It is the condition of possibility for newness—a "guide to human life" as Bachelard writes (cited in Kearney 1998, p. 7). But the imagination is not monolithic. As Alex Osborn (1993, p. 27) explains, "imagination takes many forms—some of them wild, some of them futile, some of them somewhat creative, some of them truly creative." These, Osborn continues, fall into two general categories: imagination that runs uncontrollably (dreams, hallucinations, perceptual illusions, delusions of grandeur, persecution complexes, etc.) and imagination that can be guided. Whether marked by spontaneity or by mastery, my interest here lies in imaginative acts that seed newness and not, for instance, acts of intentional recollection that can equally be seen as forms of guided imagination. From the phenomenological perspective taken here, that is, from an approach that focuses on the way we experience the imagination, the latter has three important characteristics: indeterminacy, possibility, and residuality.

Indeterminacy

Perhaps the most intuitively discovered characteristic of the imagination, aside from the relative ease in which we are able to slip into and out of imagining, is the extent to which it features indeterminacy. The imagination, in other words, is fuzzy. Some aspects of the imagined object may be envisioned vividly, accurately, and with detail, yet a degree of vagueness always persists. Edward Casey (1976), on whose phenomenological account of the imagination I rely here, describes the indeterminacy of the imagination as follows:

> no matter how much we vary imaginative experience, it always possesses a residual indeterminacy. Even if, for example, I imagine that I am in the process of *perceiving* something, what I manage to conjure up fails to possess the stolid determinateness of actually perceived objects; an intrinsic vagueness continues to characterize the imaginatively 'perceived' material. Indeed, if this vagueness were suddenly to vanish altogether from what I imagine, I would find myself no longer imagining but perceiving, remembering, or hallucinating—that is, focusing on material that is intrinsically determinate or in any case potentially determinable. (p. 110; emphasis in origin)

The objects of the imagination may appear with different degrees of clarity and texture, yet they will always suffer "a kind of essential poverty" (Sartre 2004, p. 9), lacking fidelity in comparison with perceptual data. In Jean-Paul Sartre's apt description, "there is at once too much and not enough" in the objects of imagination; "these phantom-objects are ambiguous, elusive, at once themselves and things other than themselves" (ibid., p. 132).

A similar fuzziness applies to the spatiotemporal contexts of imagined objects, "world-frames" in Casey's terms, which appear "sketchy and schematic in character, offering to the imaginer patches of space and stretches of time instead of a single coherent spatio-temporal continuum" (Casey 1976, p. 51). Furthermore, when concentrating on the liminal, nebulous, and amorphous space that surrounds the entire imaginative mise en scène, what Casey calls the "imaginal margin," the indeterminacy of imagined objects reveals itself not only as indeterminate but also as *indeterminable*: try as hard as you can and the imaginal margin will always elude specificity (ibid, p. 108). Lastly, not only is the content of the imaginative object indeterminate, but the act of imagining itself features all sorts of discontinuities, inconsistencies and lapses, stutters and jerkiness: "An imagined object does not remain present to us in an abiding manner, as do many perceptual objects; to keep it before our mental gaze, we must constantly *re*-imagine it, and even then it is difficult to say whether we are continuing to imagine exactly the *same* object again and again" (Kearney 1998, p. 7; emphasis in origin).[18]

---

[18] Sartre confirms: "the object as imaged acquires a discontinuous, jerky character: it appears, disappears, reappears and is no longer the same" (Sartre 2004, p. 135).

Possibility

While some (Sartre, for instance) have faulted the imagination precisely because it lacks determinacy, for Casey (1976) this fuzziness discloses the imagination as a site of "pure possibility":

> The indeterminacy of imaginative experience ... can be liberating to the precise extent that it does not limit or restrict what we imagine. The presence of such indeterminacy, i.e., the very lack of sharply focused detail, means that what we imagine is essentially *open* in character. (p. 110; emphasis in origin)

By its very indeterminacy, the imagination invites us to fill in the gaps and flesh out the imaginative event, object, or setting. The way we do so is essentially open—the only limits are our capacity to draw new associations between perceptive, mnemic, and imaginative objects and our willingness to do so while deferring judgment-by-reality. It is in this sense that the imagination allows us "to convert the given confines of the here and now into an open horizon of possibilities.... to anticipate how things *might be*; to envision the world *as if* it were otherwise; to make absent alternatives present to the mind's eye" (Kearney 1998, pp. 2, 6; emphasis in origin).

The imagination's capacity to open up new possibilities can also be seen as a means to reveal immanent, latent, or hidden possibilities already existing in the present—the "invisible" in Maurice Merleau-Ponty's (1968) terms. As Kearney (1998) explains, "since the invisible essence of any object can never be exhausted in a single perspectival perception, its totality can only be anticipated by means of a proleptic [pre-conceiving] imagining" (pp. 135–136). In other words, the imagination captures the world in ways that are unavailable to perception. Such a dialectical understanding of the imagination as a means to conjure presence-in-absence and absence-in-presence motivated Herbert Marcuse to hang his hopes for a new political consciousness on the capacity of the (artistic) imagination to tear the ideological veil he describes elsewhere as "one-dimensionality":

> As cognitive faculty, the imagination forms the work of art, literature, music; there it creates a reality of its own and yet *real*: in a sense more real than the given reality. Words, images, tones, gestures which deny the claim of the given reality to be all the reality and the entire reality. (Marcuse 2001, p. 117; emphasis in origin)

This "more real than real" reality is precisely that which remains hidden by one-dimensionality, a defining feature of a world in which the *given* reality is extolled as the *only* possible reality (Marcuse 1964/1991). In this sense, Marcuse hails the imagination as a means to reveal and realize the potentials tucked away by ideological, dominant forces. It is a "meta-political power" (Marcuse 2001, p. 118), a launching pad for the revolutionary consciousness.

Residuality

Seeing the imagination as a means to disclose hidden aspects of reality assumes that the imagination lies in a certain relation to reality: its objects may not be "real" in the strict sense of having an observable existence (or being available for empirical intersubjective confirmation), but neither are they unreal in the sense of lacking existence or lacking the ability to influence reality.[19] We may say, then, that imaginative objects altogether skirt the real-unreal dyad. But another way to describe this ontological lack of self-sameness is to point out that the imagination never works in a vacuum: it is closely associated with other mental faculties. A quick experiment will help demonstrate this reliance: try, if you will, to imagine a horse without appealing to previous experiences (direct or indirect) of a horse. Can you? Is it even possible? How will you know that it is indeed a horse you are imagining if you have nothing to compare it to? Imaginary objects, it follows, retain what Husserl called the "noematic nucleus" which, in this case, is derived from preexisting experience and knowledge.[20] In this context, Sartre (2004, p. 57) writes that "An image could not exist without a piece of knowledge that constitutes it," while Osborn (1993) makes a similar claim when he discusses ideation: "To develop creativeness, the mind needs not only to be exercised, but to be filled with material out of which ideas can best be formed. *The richest fuel for ideation is experience*" (p. 70; emphasis added). On these accounts, we can argue that materials furnished by other mental faculties are immanent to the imagination: past

[19] Casey (1976, p. 82) warns that to deny "ontological power to imagining" (i.e., a capability of bringing things into material existence) "is not to deny that it may be causally efficacious in certain ways."

[20] The "noematic nucleus" includes "the group of essential properties that enable us to identify a given entity, event, or state of affairs as precisely *this* entity, event, or state of affairs. It is the core-character of a phenomenon, around which cluster all secondary characteristics" (Casey 1976, p. 50, f.n. 14).

sensations, experiences, memories, and feelings persist into the imaginative content. Not only does the imagination include residual elements, its very existence relies on a measure of residuality: "The imagination is always rooted in available present and future worlds, taking its raw material, so to speak, from existing perceptions, experiences and memories. But it also reaches into the future, allowing us to project, extrapolate, surmise and speculate about things that may not exist materially, and events that have yet to take place" (Vervoort et al. 2015, p. 66).

What is extraordinary about the imagination, then, is its capacity to take these materials and refashion them into something new—to borrow from past experiences only to transcend them. John Dewey (1934) describes the synthesizing capacity of the imagination as a form of alchemy that occurs at the interface of mind and world:

> More perhaps than any other phase of the human contribution, it has been treated as a special and self-contained faculty, different from others in possession of mysterious potencies.... It is a *way* of seeing and feeling things as they compose an integral whole. It is the large and generous blending of interests at the point where the mind comes in contact with the world. When old and familiar things are made new in experience, there is imagination. When the new is created, the far and strange become the most natural inevitable things in the world. There is always some measure of adventure in the meeting of mind and universe, and this adventure is, in its measure, imagination. (p. 267; emphasis in origin)

The imagination may be free from the constraints of perception (as Casey writes in the context of "pure possibility"), but it is not entirely divorced from the vicissitudes of the phenomenological "world" (pace Sartre's objection).[21] It is how the imagination navigates these two sets of constraints that determines its capacity to propose new worldly configurations with particular form and flavor.

As we have seen above, imagined objects are never as fully formed and with as much fidelity as perceptive objects. They are inherently fuzzy, ephemeral, and contingent. This incompleteness, however, is also a sign of the imagination's immanent openness—an invitation to further fill in the

---

[21] Sartre (2004, p. 136) takes the fleetingness of imaginative objects—"they are given as perpetual 'elsewhere'"—as a sign that they manifest an "anti-world." Casey (1976, p. 51), however, disagrees, and suggests that "mini-worlds" may better describe the contextual elements of imaginative objects.

gaps by making new connections and associations. The imagination, in this sense, both requires the scaffolding offered by materials given to us by other mental faculties (perception, conception, memory, etc.), and acts itself as a scaffold for further mental activity. This double scaffolding is what allows the imagination to transcend the given, and what I take to be the real meaning of what Sartre describes as the tendency of imaginative objects to be "at once themselves and things other than themselves" (Sartre 2004, p. 132).

Unfinished Media

Ontologically self-same or not, the imagination appears essentially incomplete. This invites certain media to provide scaffolding for imaginative engagement—to perform a "structural homology" (Eco 1989, p. 18) between themselves and the imagination. What I will call here "unfinished media" do precisely that in order to engage, evoke, provoke, or stir their users' imagination.

In an interview with *Wired* magazine's Kevin Kelly, Brian Eno, the maverick musician, made an insightful comment about interactive media. After expressing his dissatisfaction with the term "interactive" as a descriptor of new media, he says:

> In a blinding flash of inspiration, the other day I realized that 'interactive' anything is the wrong word. Interactive makes you imagine people sitting with their hands on controls, some kind of gamelike thing. *The right word is 'unfinished'*. Think of cultural products, or art works, or the people who use them even, as being unfinished. *Permanently unfinished*. We come from a cultural heritage that says things have a 'nature', and that this nature is fixed and describable. We find more and more that this idea is insupportable – the 'nature' of something is not by any means singular, and depends on where and when you find it, and what you want it for. The functional identity of things is a product of our interaction with them. And our own identities are products of our interaction with everything else. (Kelly 1995; emphasis added)

Eno's comments can be seen as an indication of the way new media invite material appropriation, anticipating the rise of the internet as a site for distributed content creation. This view would later peak in popular culture with the celebration of Web 2.0 platforms and the selection of us, internet users, as *Time Magazine*'s Person of the Year in 2006 (as briefly discussed in Chap. 1). But Eno's words could also mean something deeper, reflecting the belief that neither users nor the technologies they use are self-sufficient, complete, or fixed; both are mutually evolving, entangled in ever-expanding networks of functional and semiotic value.[22] We complete our media, and our media complete us.

Is it mere chance that Eno, known for his popularizing of modern music, arrived at similar conclusions to those drawn by Umberto Eco in his important account of unfinished texts, *The Open Work*—an investigation that takes modern music as its point of departure? For Eco (1989), musical works whose score leaves significant room for the performer to enact their own interpretation of the piece with minimal guidance from the composer "appeal to the initiative of the individual performer, and hence they offer themselves not as finite works which prescribe specific repetition along given structural coordinates but as 'open' works, which are brought to their conclusion by the performer at the same time as he experiences them on an aesthetic plane" (p. 3). This kind of openness, writes Eco, is a hallmark of modern art's rejection of previous attempts to narrow down and circumscribe the range of possible interpretations of any given artwork. Although those pre-modern works could be interpreted in different ways as well—Eco writes elsewhere that every text is a "lazy machine asking the reader to do some of its work" (Eco 1994, p. 3)—and some of those interpretations may far exceed the intentions of their authors, "open" works intentionally seed different ways to interpret them. They seek to facilitate "heteroglossia" (Bakhtin 1981) by leaving considerable gaps for their readers (or performers) to fill in: "the author seems to hand them [open works] on to the performer more or less like the components of a construction kit. He seems to be unconcerned about the manner of their eventual deployment" (Eco 1989, p. 4). The openness of "open" works, therefore, is not coincidental or tangential but programmatic (Caesar 1999, p. 23). It proceeds by *signaling* its openness, *suggesting* ways in which it may be opened further, and *prompting* the reader/performer to complete or "finish" it. The temporal loop in Joyce's *Finnegan's Wake*, the lack of narrative closure in Brecht's plays, and the structural liberty of Stockhausen's musical notations are only a few examples of how this unfinishedness unfolds.

---

[22] What Stiegler (1998) calls "technogenesis" attributes this phenomenon with evolutionary weight.

A quick scan of everyday life reveals the presence of unfinished media from jigsaw puzzles and coloring books to drawing-by-numbers sketchbooks and *Choose Your Own Adventure* books. In each case, some preexisting elements serve to scaffold the imagination, at once indicating that the work is indeed open while hinting at how it may be completed. The tabs and blanks of jigsaw puzzle pieces, the illustration in the coloring book, the sequence of numbers in drawing-by-numbers books, and the available story options in *Choose Your Own Adventure* books invite the reader (used expansively here) to finish the work but to do so in one of several allowable ways. Unfinishedness, therefore, appears as a range of escalating possibilities—from the almost entirely finished to the almost entirely unfinished—with each step on the spectrum evoking a different kind of imaginative response from users (cf. McLuhan's (1964) differentiation between "hot" and "cold" media). From this perspective, *Gardens of the Anthropocene* may only be moderately unfinished, whereas my next example wears its unfinishedness on its sleeve.

IMAGINARY WORLDS (OR, "THE STUFF BEHIND SUSTAINABILITY")

Sustainability in an Imaginary World is an interactive installation that combines theatrical elements, including an actor, light projections, and elaborate sets, with several digital and analog choice-making activities.[23] It was designed by an interdisciplinary team of artists, interaction designers, and social scientists, and was open for the public for two weeks during winter 2016, and again, in a slightly modified form, for two weeks during winter 2017. The installation experimented with two ideas: first, that sustainability is premised in deeply held beliefs about the world and that attempts to engage the public with sustainability can and should address those beliefs. Second, that there are untapped communicative potentials beyond the infocentrism that characterizes many efforts to engage the public with sustainability (as discussed in Chap. 2), and that artistic ways of knowing and expressing may promote a different, more reflexive approach to sustainability.

The installation was constructed at the Centre for Interactive Research on Sustainability (CIRS), on the campus of the University of British

---

[23] I had the privilege of taking part in the installation's design and initial evaluation. The account of the installation given here refers to its first run, and draws from previously published articles (Bendor 2018; Bendor et al. 2015, 2017).

Columbia in Vancouver, Canada. At the entrance to the installation participants, in groups of six, and equipped with earphones and location-sensitive devices, were shown a brief fictitious news segment that describes a series of extreme weather events and the social and political upheavals they wrought. Once the "fake news" segment ends, an actor dressed as a janitor enters the room and hints at an unfolding conspiracy to mask what is really happening. He urges participants to check it out for themselves, and ushers them into the installation's main gallery, a dark, industrial space sparsely populated with wooden crates. In the main gallery, participants can hear three distinct voices, each describing and interpreting the events featured in the newscast from their own perspective. After one of the crates reveals itself as a touchscreen table (see Image 5.2), participants are asked to make two choices about how society should address the unfolding climate crisis: should technological innovation or changes in lifestyle drive societal responses? And should responses be based on collective solutions or on individual action? Aggregated answers are represented by a symbolic image that serves as a visual through line, accompanying the group as it moves through the installation.

Image 5.2   Group decision-making on a touchscreen table in *Sustainability in an Imaginary World*. (Image by Emily Cooper)

Image 5.3 Light cues indicate the presence of three additional rooms in *Sustainability in an Imaginary World*. (Image by Emily Cooper)

Once decision-making is complete, the installation turns much more abstract. Three faint, door-shaped projections on the main gallery's walls indicate the existence of additional spaces (see Image 5.3), yet participants are not given any other cues or further instructions. They are free to explore the rooms in groups or alone, in the sequence of their choice, or not at all.

The three rooms that open up from the main gallery space manifest three different "worlds"—spiritual, materialist, and literary.[24] They include light and sound effects, a representation of a tree, color-coded leaves, and various interactive elements. The spiritual room is blanched and features two trees—one hanging upside down, branches touching—encircled by a narrow path that, along with the masks that hang on the wall, suggest ritualistic possibilities (see Image 5.4). The materialist room is designed with steam-punk aesthetic. It is copper colored and features a metallic tree with light bulbs, wires, pulleys, locks, and keys (see Image 5.5). If participants manage to open the locks and pull on the appropriate strings, the

[24] The choice of worlds was inspired by Rorty (2007).

Image 5.4   The "spiritual" room in *Sustainability in an Imaginary World.* (Image by Emily Cooper)

tree blossoms with paper flowers. Lastly, the literary room features nine doors that open to all sorts of funhouse mirrors, curiosity cabinets, live closed-circuit camera feeds, and a tree segmented into glass jars (see Image 5.6). Green leaves are spread around the room.

After participants have had the opportunity to explore and interact with the three rooms, they are hurried back to the installation's main gallery by the janitor. But just before leaving the space, a crate descends from the ceiling and a voice invites participants to make one final choice by plucking a colored leaf from the crate's surface: "White leaves for the knowledge that something is out there. Copper leaves for the faith that answers will come. Green leaves for the comfort in knowing no world is carved in stone." With every plucked leaf, same-colored trees appear on the main gallery's walls, transforming the room into a tricolored forest that represents the group's cumulative choices.

Image 5.5 The "materialist" room in *Sustainability in an Imaginary World*. (Image by Emily Cooper)

The installation is rich in symbolism and meaning. It uses artistic techniques and vocabularies to convey the sense that "the ways in which we perceive, give meaning, and act on sustainability are premised in deeper cognitive, cultural, and ontological structures," and therefore, "How we understand sustainability … is inherently tied to how we understand the world, ourselves, and others, and to the ways in which we render those beliefs actionable" (Bendor et al. 2017, p. 6). Furthermore, the installation suggests that these "worlds" can co-exist, that is, that reality is plural; we live in a pluriverse. Not only do we inhabit multiple worlds, but we can choose among them, create new ones, and engage in constructive dialogue with others about the merits and disadvantages of different worlds (see Goodman 1978; Vervoort et al. 2015). From such a social constructivist perspective, the installation puts forth the notion that "we are all worldmakers" (Bendor 2018).

The installation's message about the plurality of worlds and the consequent plurality of ways to interpret and act on sustainability finds a suitable homology in its form—in the way it performs its unfinishedness by making use of playful ambiguity and open-endedness. Participants are not given

Image 5.6   The "literary" room in *Sustainability in an Imaginary World*. (Image by Emily Cooper)

explicit instructions about how to move through the installation or inter-act with its different features, yet without their active participation, the installation remains mute. Furthermore, participants are not forced to interpret the installation in any particular way, and are only given partial clues as to how the different elements may fit together into a coherent whole. Participants, in other words, face the real possibility of leaving the installation puzzled, perplexed, or even frustrated (see Bendor et al. 2017). Whenever participants encounter a familiar trope or activity, they soon realize that it does not necessarily indicate how the rest of the experience will unfold: the initial narrative presents a fairly "standard" cli-fi (climate fiction) situation only to plunge participants into a much more abstract sequence of experiences; choice-making on the touchscreen table retrieves the familiar trade-offs associated with sustainability, yet the results do not alter the rest of the experience; "busy" interactive features that indicate that the installation is perhaps a solvable puzzle lead to moments of slow, actionless contemplation. The installation, in other words, does not pro-vide a complete "story" but serves as a scaffold—presenting participants with "blank canvases" on which to "project their own imagination, while providing them with enough of a through line to be able to follow the experience to its conclusion" (Bendor 2018). And while initial results indicated that not all visitors to the installation experienced the kind of interpretive openness and agentic possibility intended by the designers (Bendor et al. 2017), some visitors have. Take, for example, the account given by a visitor after being asked about his interpretation of the installation:

> The multitude of choices that we have with regard to our collective future and sustainability, I thought that did a really good job of capturing the way that can feel in action.... I thought the rooms were very dreamlike, introducing ways out, things that society could do ... we don't know for sure what's going to happen and we don't know for sure what the choices are going to feel like when they're put into practice. Yes, it is about hypotheses and choosing a path forward and owning it and wanting to do it. I felt like it was the stuff behind sustainability ...[25]

At least for some visitors, glimpsing "the stuff behind sustainability" opened up ways to reflect on sustainability as a repository of potential meanings and actions—as an imaginary.

So far we have discussed several ways in which imaginaries may be impacted. *Life 2053*, the Swedish online exhibition, illustrated how futuristic sustainable lifestyles may inspire a sense of achievability, although, it was argued, the exhibition does not go as far as it could in terms of proposing an alternative social imaginary. *Gardens of the Anthropocene* demonstrated how alternative scientific imaginaries may be brought to life as interactive environments, raising provocative questions about floral futures as a way to critically point to the boundary work upon which modernity was established. *Sustainability in an Imaginary World* illustrated how unfinished media may provide users scaffolding for imaginative engagement with the ontological premises of sustainability. Playful ambiguity and open-endedness were discussed as means to evoke participants' imagination without prefiguring or constraining the range of interpretations at participants' disposal. The next example illustrates a different approach. Instead of inspiring, challenging, or scaffolding the imagination, this approach lets users rehearse alternative imaginaries—imagine a better future and "try it on for size."

---

[25] Transcribed from a group interview following their visit to installation on January 29, 2016.

A New Political Imaginary?

Fort McMoney is an interactive documentary game that takes players on a journey into Fort McMurray, Alberta, the notorious tar-sands boomtown, and asks them to make a series of decisions about the town's future. The result of two years of research, the game was created by David Dufresne and co-produced by Toxa and the National Film Board of Canada, in collaboration with the Franco-German studio Arte.[26] It can be played in English, French, or German, and in the first eight months since its launch at the end of 2013, it has attracted more than 400,000 players from around the world (Uricchio et al. 2015, p. 100).

The narrator introduces the game to players during the first "episode":

> You have reached the end of the road, at the world's edge. Fort McMurray, an area the size of Florida, larger than Hungary. You are embarking on a documentary game in which everything is real—the places, the events, the characters. The choices you will make will determine your experience and will affect the other players in the game. You will be initiating encounters and collecting clues. Your mission: to visit Fort McMurray, measure what's at stake, vote on referendums, and debate with other players, and find a way to achieve the impossible … Fort McMoney's fate is in your hands.

During the game's four-week duration (consisting of three weekly episodes and one week for deliberation), players learn about the town, debate the issues and possible solutions with other players, and make a series of decisions about the town's future. Players can watch more than 100 short videos including 55 interviews with real characters, from exotic dancers to the town's mayor, business owners, and aboriginal leaders. They can learn more about what brought those people to the town and get a first-person perspective on the issues with which they struggle. Players can also go on collaborative missions (like crossing an ice bridge) that give them "influence points" that can be used later. All this activity is managed from a central control panel that provides players with access to news, online discussions, the results of previous and ongoing voting, and more.

Much has been said about *Fort McMoney*'s unique format—about its relation to digital journalism (Uricchio et al. 2015) or to documentary

---

[26] *Fort McMoney* is available online at http://www.fortmcmoney.com/#/fortmcmone (last accessed Mar. 18, 2018), but live discussions are no longer active.

films more generally (Nash 2014; Nogueira 2015). But in the context of sustainability, the game conveys a real sense of the struggles involved in pursuing sustainability in the age of fossil capital. In Dufresne's words, "Capitalism is the biggest game in the world … Everyone plays it every day. And Fort McMurray is the most capitalist city in the world. That's why we call it 'Fort McMoney'" (cited in Goldberg 2013, Nov. 26). There is little doubt that Fort McMurray, the actually existing town, has benefited financially from the tar sands' attraction of newcomers. But rapid, all too rapid changes also created a series of tensions around the distribution of wealth and responsibilities in the town: how can the "frontier mentality" of oil patch workers be kept in check by the rule of law? How can the economic benefits of fossil fuel extraction be measured against the latter's environmental costs? And how can residents maintain and distribute fairly the town's growing riches in the future? These questions and the social dilemmas they unfold are brought to life by the game's interactive features. Players experience the town firsthand; real people and real stories bring depth to what is often debated in mainstream media with simplified socioeconomic abstractions. In *Fort McMoney*, as the game guide puts it, "Nothing is black and white, and solutions are not obvious. Navigating the grey zones is the key to understanding what's at stake when it comes to oil extraction" (Wohlberg 2013, Dec. 2). This, of course, is not to say that the game maintains a form of (idealized, untenable) neutrality but that it refuses to frame the issues in a cut-and-dry manner. The Fort McMurray that emerges from *Fort McMoney* is "a city of complexities" (Dufresne, quoted in Goldberg 2013, Nov. 26).

The game, however, does more than provide an intimate look into the trade-offs that are so often associated with sustainability (the familiar jobs vs. environment trope, for instance), or humanize those that benefit from the fossil fuel industry at the expense of the planet. The human consequences of Alberta's oil rush, the heroes, villains, and casualties of a culture dominated by short-termism and the pursuit of financial gain at all costs become the backdrop for what is essentially an exercise in deliberative democracy. The staggeringly complex set of issues faced by the town, the degree to which the town must balance its residents' desires and frailties with concerns about social, economic, and environmental justice, make the town a perfect lab for experimenting with the kind of rational democratic sensibilities advocated by Habermas (and as discussed in Chap. 3). The game, in other words, creates a quasi-deliberative democracy by its very affordances, promoting the exchange of informed opinion as a means to arrive at political decisions. The decisions made in the game may not have "real-world" consequences insofar as they are not legally nor otherwise binding for Fort McMurray, but the opportunity to engage in political deliberation can be both liberating and insightful. Participants in the early stages of the game have voted, for instance, to increase taxation on petroleum products, to tighten environmental controls over extraction, and even to nationalize the oil industry altogether (a decision that was supported by nearly 500,000 votes, over 75% of players!).[27] Does such a decision merely reveal the biases or predispositions of players? Could it have been reached in the "real world" where its impact would be felt in a very nonvirtual way? It is hard to say. But if we agree with Klein (2014) that social change starts with the ability to imagine the world differently, the game's capacity to let players consider, debate, and "try for size" sustainability-related decision-making should not be understated. Something else that should not be downplayed is the game's capacity to build a sense of collectivity and solidarity: "a great deal of the work of deep social change involves having debates during which new stories can be told to replace the ones that have failed us" (Klein 2014, p. 461). The freedom to navigate the game's narrative and environment and to participate in free-formed discussions invites players to tell such new stories about Fort McMurray and to "imagine an alternative Fort McMurray" (Nogueira 2015, p. 91). Even if the real Fort McMurray may not be affected, the lessons of *Fort McMoney* are likely to carry over into the real world. One can only wonder whether those who have discounted their own capacity to affect political decisions about oil tanker traffic on BC's coast would have answered the poll differently had they played a similar game.

Escape Velocity

Writing about the imagination's role in producing climate change as a cultural object, Kathryn Yusoff and Jennifer Gabrys (2011) argue that "The work of the imagination is a will to become; in many different ways the imagination extends, pushes, challenges, and confides to us what the human is. But, perhaps the greatest work of the imagination is its counterweight to

---

[27]'Should oil be nationalized? Fort McMoney players say "yes"'. *The Globe and Mail*. Retrieved from https://beta.theglobeandmail.com/report-on-business/industry-news/energy-and-resources/should-oil-be-nationalized/article15809881 (last accessed Mar. 18, 2018).

the actuality of the world, to imagine how we might be otherwise" (p. 529). The imagination is a means for breaking the seductive yet nefarious hegemonic view of the given as the only possible reality—to achieve the velocity necessary to escape the gravitational pull of the here and now. In Richard Rorty's (2007, p. 13) words, "The only source of redemption is the human imagination." Since we must be able to imagine change before we can pursue it, interactive media, as illustrated above, can support the transformational capacity of the imagination in several ways.

First, interactive media may inspire new imaginaries by providing users with examples of existing or nearly-existing alternative social, political, economic, or cultural configurations. Seeing, or, better yet, experiencing such alternatives signals to the imagination that the given world is only one of many possibilities (this is one way in which the resonant interactions discussed in the previous chapter could compliment worldmaking interactions). As individuals and as a collective, we encounter bifurcation points on every step, and the choices we make are situated and thus highly contingent. The Swedish online exhibition *Life 2053* illustrates this, but it also exemplifies how focusing on concrete, achievable alternatives may dilute or even counteract opportunities for more radical alternatives. The exhibition proposes that technological development may make available more environmentally benign lifestyles, but it does not challenge the values, worldviews, and assumptions that ground those lifestyles. Is it reasonable to expect that in 30 years or so, and after witnessing the effects of climate change even more up close, we will still hold the same kind of consumerist values? Inspiring a sense of possibility may indeed help shift awareness into action and, in this sense, help bolster self-efficacy, but does it challenge our imaginaries in meaningful ways? Does it amount to anything more than "solutionism" (Morozov 2013, Mar. 2)?

As *Life 2053* illustrates, some interactive media seek to inspire the imagination and extend existing imaginaries. Other media, however, launch a more direct challenge to existing imaginaries. They suggest new imaginaries not as a means for identifying or forecasting possible futures but as a method to radically challenge existing ones. As such, they manifest a form of speculative design (Dunne and Raby 2013), using possible futures "as tools to better understand the present" (ibid., p. 2). Matt Malpass (2017) adds that "Rather than presenting utopic or dystopian visions, speculative designs pose challenging statements that attempt to explore ethical and societal implications of new science ... The aim is to make scientific theories and the cultural implications of science perceptible in different ways"

(p. 101). *Gardens of the Anthropocene* does exactly that by extrapolating from existing biological knowledge a world in which new organisms transgress the nature-culture divide. The (augmented) reality it animates problematizes our existing ways of knowing and being, and in this sense, the installation manifests an Anthropocene like we have yet to experience. The way the imaginary flora is superimposed on the existing environment provides the installation with a sense of realism, yet it does so without claiming reality. Unlike the media discussed in the previous chapter, *Gardens* does not provide users with a complete, diegetically consistent world to inhabit but only with scaffolds. The installation is a form of unfinished media: it invites users to fill in with their own imagination the blanks left by the installation's form.

Identifying unfinishedness as an imagination-provoking strategy leads to a third way in which worldmaking interactions may unfold. If *Gardens* uses art to challenge scientific imaginaries, *Sustainability an in Imaginary World*, the interactive theatrical installation, reaches deeper into the artistic bag of tricks to pluralize and multiply worlds. It makes use of playful ambiguity and open-endedness to undermine the relatively stable regime of meaning carried by any single world. The plurality and co-existence of worlds (spiritual, materialistic, and literary, in the terms used by the installation) is given tangible form in a homologous hermeneutic: a plurality of worlds translates to a plurality of meanings and a plurality of strategies for meaning-making. Instead of undermining existing imaginaries or proposing preferable, alternative ones, the installation suggests that not only are imaginaries malleable by their very essence, but that anyone can imagine the world and give it form and meaning in multiple, equally valuable ways—individually and collectively.

The last way in which worldmaking interactions may generate escape velocity is less ontological and more practical. Here, alternative imaginaries are rehearsed, performed, and tried out for size. The interactive documentary game *Fort McMoney* was designed to inform players about the everyday lives and challenges of those living in areas dominated by fossil fuel extraction, thus providing players with an entry point into the complex reality and the stakes involved in extractive economies. The physical setting of the game, Fort McMurray, is a real town with real problems—problems made all the more real as they are discussed by the town's inhabitants on location. Giving players a firsthand, intimate view of the challenges faced by the town, however, serves the game's other objective: providing players with an opportunity to rehearse deliberative democratic procedures. As players watch short informational videos, debate the issues, and vote on different questions, they get a sense of what it would be like to live in a world where such things are not considered political luxury but part of everyday life.

Inspiring, challenging, multiplying, and rehearsing alternative imaginaries are techniques for evoking the imagination, but the outcomes of imaginative engagement need to be translated into collective possibilities for change. Worldmaking interactions, as they emerge from this chapter, are a promising means to do so—not only by promoting creativity, an individual's capacity to generate newness, but by channeling creativity into social strategies of meaning-making, visioning, and pursuing change. Through worldmaking interactions, the imagination finds an object to transform. Much like the "mattering maps" discussed in the previous chapter, worldmaking interactions provide social meaning and outlets to individual capacities. In *Life 2053* that is done by linking individual lifestyle choices to global material outcomes. In *Gardens of the Anthropocene*, this unfolds by giving scientific imaginaries a localized, experienceable presence. *Sustainability in an imaginary World* makes experiential the relations between individual meaning-making strategies and ontological versions of the world. And *Fort McMoney* performs an idealized accumulation of individual preferences into new, collective political visions. In each case, interactive media are deployed as a means to consummate a sense of possibility not for the sake of curiosity or creativity but as part of a tangible political agenda. They compel us to ask, "if things could be different, why are they still the same?" Achieving escape velocity, then, is about more than finding creative solutions to existing problems; it is about envisioning and pursuing what Deleuze and Guattari (1987) call "lines of flight", generating momentum for radical transformation. Recognizing the possibility of otherness gives one faith that change is possible, and hope that it may indeed come about. More on that in the final chapter.

BIBLIOGRAPHY

Alvarez, A. C. (2017, April 26). Envisioning New Futures: Steve Lambert and Stephen Duncombe on Artistic Activism. *Creative Capital Blog*. Retrieved from http://blog.creative-capital.org/2017/04/envisionng-new-futures-steve-lambert-stephen-duncombe-artistic-activism/

Arias-Maldonado, M. (2013). Rethinking Sustainability in the Anthropocene. *Environmental Politics, 22*(3), 428–446.

Augé, M. (2015). *The Future*. London/New York: Verso.

Bakhtin, M. M. (1981). *The Dialogic Imagination: Four Essays* (trans: Holquist, M.). Austin: University of Texas Press.

Bendor, R. (2018). Interaction Design for Sustainability Futures: Towards Worldmaking Interactions. In M. Hazas & L. P. Nathan (Eds.), *Digital Technology and Sustainability: Engaging the Paradox* (pp. 205–216). New York: Routledge.

Bendor, R., Anacleto, J., Facey, D., Fels, S., Herron, T., Maggs, D., et al. (2015). Sustainability in an Imaginary World. *Interactions, 22*(5), 54–57.

Bendor, R., Maggs, D., Peake, R., Robinson, J., & Williams, S. (2017). The Imaginary Worlds of Sustainability: Observations from an Interactive Art Installation. *Ecology and Society, 22*(2), 17.

Bonneuil, C., & Fressoz, J.-B. (2017). *The Shock of the Anthropocene*. London/New York: Verso.

Bottici, C. (2014). *Imaginal Politics: Images Beyond Imagination and the Imaginary*. New York: Columbia University Press.

Caesar, M. (1999). *Umberto Eco: Philosophy, Semiotics and the Work of Fiction*. Oxford/Malden, MA: Polity Press.

Caprara, G. V., Vecchione, M., Capanna, C., & Mebane, M. (2009). Perceived Political Self-Efficacy: Theory, Assessment, and Applications. *European Journal of Social Psychology, 39*(6), 1002–1020.

Casey, E. S. (1976). *Imagining: A Phenomenological Study*. Bloomington: Indiana University Press.

Castoriadis, C. (1997). *The Imaginary Institution of Society* (trans: Blamey, K.). Cambridge, MA: MIT Press.

Cross, K., Gunster, S., Piotrowski, M., & Daub, S. (2015). *News Media and Climate Politics: Civic Engagement and Political Efficacy in a Climate of Reluctant Cynicism*. Vancouver: Canadian Centre for Policy Alternatives. Retrieved from https://www.policyalternatives.ca/sites/default/files/uploads/publications/BC_Office/2015/09/CCPA-BC-News_Media_Climate_Politics.pdf

Deleuze, G., & Guattari, F. (1987). *A Thousand Plateaus: Capitalism and Schizophrenia* (trans: Massumi, B.). Minneapolis: University of Minnesota Press.

Dewey, J. (1934). *Art as Experience*. New York: Paragon.

Dijksterhuis, E. J. (1961). *The Mechanization of the World Picture* (trans: Dikshoorn, C.). London/New York: Oxford University Press.

Dunne, A., & Raby, F. (2013). *Speculative Everything: Design, Fiction, and Social Dreaming*. Cambridge, MA: MIT Press.

Eco, U. (1989). *The Open Work* (trans: Cancogni, A.). Cambridge, MA: Harvard University Press.

Eco, U. (1994). *Six Walks in the Fictional Woods*. Cambridge, MA: Harvard University Press.

Ehgartner, U., Gould, P., & Hudson, M. (2017). On the Obsolescence of Human Beings in Sustainable Development. *Global Discourse, 7*(1), 66–83.

Ehrenfeld, J. (2008). *Sustainability by Design: A Subversive Strategy for Transforming Our Consumer Culture*. New Haven: Yale University Press.

Eisenstein, C. (2011). *Sacred Economics: Money, Gift, and Society in the Age of Transition*. Berkeley: North Atlantic Books.

Erikson, J. (2012, March 5). Overcoming Barriers to a Green Economy. *SustainAbility*. Retrieved from http://sustainability.com/ourwork/insights/overcoming-barriers-to-a-green-economy/

Freud, S. (1920). *A General Introduction to Psychoanalysis* (trans: Hall, G. S.). New York: Boni & Liveright Publishers.

Garrard, G., Handwerk, G., & Wilke, S. (2014). Introduction: "Imagining Anew: Challenges of Representing the Anthropocene". *Environmental Humanities, 5*, 149–153.

Ghosh, A. (2016, October 28). Where Is the Fiction About Climate Change? *The Guardian*. Retrieved from https://www.theguardian.com/books/2016/oct/28/amitav-ghosh-where-is-the-fiction-about-climate-change-_-__jmp0_

Goldberg, H. (2013, November 26). Where Film Marries Video Game. *New York Times*. Retrieved from http://www.nytimes.com/2013/11/27/arts/video-games/where-film-marries-video-game.html

Goodman, N. (1978). *Ways of Worldmaking*. Indianapolis: Hackett Pub. Co.

Haiven, M. (2014). *Crises of Imagination, Crises of Power: Capitalism, Creativity and the Commons*. London: Zed Books.

Haiven, M., & Khasnabish, A. (2014). *The Radical Imagination: Social Movement Research in the Age of Austerity*. London: Zed Books.

Kearney, R. (1998). *Poetics of Imagining: Modern to Post-Modern* (2nd ed.). New York: Fordham University Press.

Kelly, K. (1995, June). Gossip Is Philosophy. Kevin Kelly Talks to Brian Eno. *Wired, 3.05*, 146–151, 204–149.

Klein, N. (2014). *This Changes Everything: Capitalism vs. the Climate*. Toronto: Alfred A. Knopf.

Kuhn, T. S. (1962). *The Structure of Scientific Revolutions*. Chicago: University of Chicago Press.

Latour, B. (2014). Agency at the Time of the Anthropocene. *New Literary History, 45*(1), 1–18.

Latour, B. (2017). *Facing Gaia: Eight Lectures on the New Climatic Regime* (trans: Porter, C.). Cambridge/Medford: Polity.

Lövbrand, E., Beck, S., Chilvers, J., Forsyth, T., Hedrén, J., Hulme, M., et al. (2015). Who Speaks for the Future of Earth? How Critical Social Science Can Extend the Conversation on the Anthropocene. *Global Environmental Change, 32*, 211–218.

Maggs, D., & Robinson, J. (2016). Recalibrating the Anthropocene: Sustainability in an Imaginary World. *Environmental Philosophy, 13*(2), 175–194.

Malm, A. (2016). *Fossil Capital: The Rise of Steam-Power and the Roots of Global Warming*. London/New York: Verso.

Malpass, M. (2017). *Critical Design in Context: History, Theory, and Practices*. London/New York: Bloomsburry.

Mann, G., & Wainwright, J. (2018). *Climate Leviathan: A Political Theory of Our Planetary Future*. London/New York: Verso.

Marcuse, H. (1964/1991). One-Dimensional Man: Studies in the Ideology of Advanced Industrial Society. Boston: Beacon Press.

Marcuse, H. (2001). Beyond One-Dimensional Man. In D. Kellner (Ed.), *Towards a Critical Theory of Society* (pp. 111–120). London/New York: Routledge.

McLuhan, M. (1964). *Understanding Media; The Extensions of Man*. New York: McGraw-Hill.

McNichol, T. (2010). The Art Museum as Laboratory for Reimagining a Sustainable Future. In T. Thatchenkery, D. L. Cooperrider, & M. Avital (Eds.), *Positive Design and Appreciative Construction: From Sustainable Development to Sustainable Value* (pp. 177–193). Bingley: Emerald Group Publishing Limited.

Merleau-Ponty, M. (1968). *The Visible and the Invisible* (trans: Lingis, A.). Evanston: Northwestern University Press.

Monbiot, G. (2017, September 9). How Do We Get Out of This Mess? *The Guardian*. Retrieved from https://www.theguardian.com/books/2017/sep/09/george-monbiot-how-de-we-get-out-of-this-mess

Morozov, E. (2013, March 2). The Perils of Perfection. *New York Times*. Retrieved from http://www.nytimes.com/2013/03/03/opinion/sunday/the-perils-of-perfection.html

Nash, K. (2014). Clicking on the Real: Telling Stories and Engaging Audiences Through Interactive Documentaries. Retrieved from http://eprints.lse.ac.uk/71175/1/blogs.lse.ac.uk-Clicking%20on%20the%20real%20telling%20stories%20and%20engaging%20audiences%20through%20interactive%20documentaries.pdf

Nogueira, P. (2015). Ways of Feeling: Audience's Meaning Making in Interactive Documentary Through an Analysis of Fort McMoney. *Punctum, 1*(1), 79–93.

Osborn, A. F. (1993). *Applied Imagination; Principles and Procedures of Creative Problem-Solving* (3rd rev. ed.). Buffalo: Creative Education Foundation.

Rorty, R. (2007). Philosophy as a Transitional Genre. In *Philosophy as Cultural Politics* (pp. 3–28). Cambridge: Cambridge University Press.

Sartre, J.-P. (2004). *The Imaginary: A Phenomenological Psychology of the Imagination* (trans: Webber, J.). London/New York: Routledge.

Stiegler, B. (1998). *Technics and Time (Vol. 1: The Fault of Epimetheus)* (trans: Beardsworth, R.). Stanford: Stanford University Press.

Taylor, C. (2004). *Modern Social Imaginaries*. Durham: Duke University Press.

Tresch, J. (2007). Technological World-Pictures: Cosmic Things and Cosmograms. *Isis, 98*(1), 84–99.

Trischler, H. (2016). The Anthropocene: A Challenge for the History of Science, Technology, and the Environment. *NTM, 24*(24), 309–335.

Turner, C. (2007). *The Geography of Hope: A Tour of the World We Need*. Toronto: Random House Canada.

Uricchio, W., Wolozin, S., Bui, L., Flynn, S., & Tortum, D. (2015). *Mapping the Intersection of Two Cultures: Interactive Documentary and Digital Journalism*. Retrieved from http://opendoclab.mit.edu/interactivejournalism/

Vervoort, J. M., Bendor, R., Kelliher, A., Strik, O., & Helfgott, A. E. R. (2015). Scenarios and the Art of Worldmaking. *Futures, 74*, 62–70.

Wals, A. E. J., & Corcoran, P. B. (2012). Re-orienting, Re-connecting and Re-imagining: Learning-Based Responses to the Challenge of (Un)sustainability. In A. E. J. Wals & P. B. Corcoran (Eds.), *Learning for Sustainability in Times of Accelerating Change* (pp. 21–32). Wageningen: Wageningen Academic Publishers.

Williams, A., & Srnicek, N. (2013). # Accelerate Manifesto for an Accelerationist Politics. *Critical Legal Thinking, 14*, 72–98.

Wohlberg, M. (2013, December 2). Ideas Battle Online for Control of Fort McMoney. *Northern Journal*. Retrieved from https://norj.ca/2013/12/ideas-battle-online-for-control-of-fort-mcmoney/

Wright, C., Nyberg, D., De Cock, C., & Whiteman, G. (2013). Future Imaginings: Organizing in Response to Climate Change. *Organization, 20*(5), 647–658.

Yusoff, K., & Gabrys, J. (2011). Climate Change and the Imagination. *Wiley Interdisciplinary Reviews: Climate Change, 2*, 516–534.

CHAPTER 6

Refractions

A Contested Concept

Despite its presence for more than 200 years—at least since the 1700s writes Caradonna (2014), even longer suggests Grober (2012)—sustainability's translation into a concrete sociopolitical program has not been smooth nor straightforward. While gaining traction and popularity, the term has also maintained a certain degree of abstractedness. Sustainability's definitional fuzziness is often traced back to the Brundtland Commission's report *Our Common Future* (1987), where sustainability (then "sustainable development") was defined as "development that meets the needs of the present without compromising the ability of future generations to meet their own needs" (UNWCED 1987, ch.2). While the report establishes strong links between environmental degradation and socioeconomic inequality, it leaves open the question of what constitutes "needs" or how they may be fulfilled. As Marshall and Toffel (2005) point out: "How should this definition be used to evaluate policy choices or business decisions? To avoid impeding the 'ability of future generations to meet their own needs' requires predicting both their needs and their abilities, which in turn requires forecasting their available technologies" (p. 673). The plethora of work that has since attempted to give the Brundtland Commission's definition more exact meaning testifies to the difficulty in answering these questions in a single, agreeable manner. "Sustainable development" has since been replaced by "sustainability," resolving the

© The Author(s) 2018
R. Bendor, *Interactive Media for Sustainability*,
Palgrave Studies in Media and Environmental Communication,
https://doi.org/10.1007/978-3-319-70383-1_6

internal tension that plagued the original term—between the perpetual drive for growth that underlies modern economies and the immutable physical limits of the planet—and in this sense acknowledging the truth of Herman Daly's (1993, p. 267) famous remark that "sustainable growth is impossible." Yet, the definitional nebulosity and implicit circularity that was the inheritance of the Brundtland Commission's definition persists.

Steve Connelly (2007) observes that responses to the vagueness of the term have taken one of four forms. The first is to simply ignore the complexity of sustainability in order to reach an operationalizable definition. Many of the persuasive interactions discussed in Chap. 2 seem to do just that when they reduce sustainability to a set of measurable and achievable material indicators. The second response acknowledges the ambiguity of the term yet proceeds to offer a specific definition, sometimes justified by tracing it back to the founding definition given by the Brundtland Commission. This approach is reflected by the resonant interactions that were discussed in Chap. 4, given the way such interactions convey sustainability's complexity through a single, felt dimension. The third response makes the term's ambiguity explicit, while proceeding to characterize different interpretations based on one or more analytical axes. The way synoptic interactions (discussed in Chap. 3) unpack complex systems based on selected factors is consonant with this type of approach. The fourth response, which Connelly advocates, understands different interpretations of sustainability as competing rhetorical claims. Responses of this type, Connelly argues, "acknowledge the intellectual legitimacy of alternative interpretations of the concept, in order to appreciate how and why they can be strongly held and defended—an acknowledgement hampered by approaches that insist that alternatives are not just undesirable (perhaps politically illegitimate) but definitionally incorrect" (ibid., p. 262). To some extent, this fourth approach is evident in the worldmaking interactions discussed in Chap. 5, but it can also be seen in recent discussions of sustainability as an "essentially contested concept" (Connelly 2007; Ehrenfeld 2008; Jacobs 1999). According to this view, sustainability inherently defies definitional precision and clarity, and as such, it exhibits the characteristics of what Ian Hacking (1999) calls an "elevator word": a word or term that is often defined circularly and fluidly, conjuring other abstract terms, which, in turn, refer to more abstract terms, and so forth. This circularity is apparent, for instance, in John Ehrenfeld's (2008) definition of sustainability as *"the possibility that humans and other life will flourish on the Earth forever*" (p. 53; emphasis in origin). Ehrenfeld's attempt to thicken the definition by adding that it includes three overlapping domains—"Our sense of ourselves as human beings: the human domain"; "Our sense of our place in the (natural) world: the natural domain"; and "Our sense of doing the right thing: the ethical domain" (ibid., p. 58)—does little to clarify what flourishing means without recourse to other, equally nebulous or contested terms (a fact Ehrenfeld himself readily admits).[1] The important point, however, is not that sustainability is inherently vague and open to interpretation, a site of semantic juggling or backflipping, but that different interpretations of sustainability are indicative of the social and political contexts within which sustainability is invoked. Different interpretations of sustainability, therefore, make evident the discursive nature of sustainability (Dryzek 2013; Jacobs 1999; Robinson 2004).

Take, for example, what Miller (2013) identifies as the two dominant approaches to sustainability: the "universalist" and the "procedural." The universalist approach aims to develop criteria for evaluating the degree to which different situations are sustainable. Such criteria are exclusively based on scientific, "universal" knowledge, and are clearly beneficial to those seeking to evaluate sustainable plans, decisions, or outcomes. Johan Rockström and colleagues' work on "planetary boundaries" (Rockström and Klum 2015; Rockström et al. 2009) and reports produced by large international organizations (see, for instance, UN 2015, p. 5) are good examples of such efforts. The universalist approach has also influenced those in the SHCI community working from a "computing within limits" perspective (Nardi et al., forthcoming). The "procedural" approach to sustainability, in contrast, shifts the focus from sustainability as an end state, evaluated against scientifically derived criteria, to the process by which sustainability is articulated—"how sustainability comes to be defined and how pathways are developed to pursue it" (Miller 2013, p. 284). As articulated by John Robinson, the procedural approach posits sustainability as a dynamic process of future-making, "the emergent property of a conversation about desired futures that is informed by some understanding of the ecological, social and economic consequences of different courses of action" (Robinson 2004, p. 381). If sustainability is a

---

[1] "It doesn't say much about how to get there and it doesn't say how we will ever know that we are indeed there" (Ehrenfeld 2008, pp. 53–54).

future-making process in which procedures and goals are "co-produced" by relevant stakeholders, it is quite reasonable to expect very different outcomes from processes that take place in different locations, with different stakeholders, and under different circumstances. What may be deemed "sustainable" for one community may not seem so sustainable to another, the upshot of which is that "sustainability emerges as a normative ethical principle rather than a scientific concept" (Maggs and Robinson 2016, p. 185).

The two approaches clearly differ on the question of whether sustainability is an end state or a process. But if we regard them as contesting positions on a discursive field, another difference emerges. The universalist approach, as Miller (2013) points out, attempts to separate the technical criteria for sustainability from its normative implications. While the first are provided by science, the second are left for the public to debate. Accordingly, "By embracing a universal sustainability, scientists are able to avoid opening up an arena in which the role of science and the knowledge produced by scientists may be contested along with other components of sustainability" (ibid., p. 284). From the perspective of the procedural approach, the entanglement of scientific knowledge and normative evaluations, "objective" facts and "subjective" values, is not only unavoidable but welcome. In this, the procedural approach reflects the growing influence of "mode 2" (Gibbons et al. 1994) or "post-normal" science (Funtowicz and Ravetz 1993), and the concomitant shift from research in the lab to "research in the wild" (Callon et al. 2009). It becomes clear, then, why the universalist approach is championed by scientists and bureaucrats: for those working from the universalist approach, scientific knowledge becomes a tool of power, capable of influencing other discursive stakeholders. But for those working from the procedural approach, scientific knowledge is only one of many possible rationalities at play in social processes of decision-making and future-making; not a form of pure, unbiased, objective knowledge but one of several legitimate perspectives. Either way, what is at stake in the "competition" between the universalist and the procedural approaches is not the meaning of sustainability proper, but the way sustainability lends itself to a diagram of knowledge/ power relations (to use Foucault's terms), according to which different "bits of information" about sustainability are put together "into coherent stories or accounts" (Dryzek 2013, p. 9).

If we accept the view of sustainability as a discursive field, two conclusions emerge. First, the meaning of sustainability does not need to be saved nor preserved (pace Marshall and Toffel 2005, p. 679), nor has it really been "hijacked" (as argued by Mittler 2001 and Parr 2009) in the first place. The definitional fuzziness is not an afterthought or blemish to remove, nor is it a strategic advantage in the "marketplace of ideas" (Caradonna 2014, p. 7), but an essential dimension of sustainability. It is a sign of discursive vitality. So while a measure of definitional fuzziness may deter those who require a precise definition from which to derive quantitative indicators and criteria, open-endedness, as discussed in Chap. 5, invites reflection and creative interpretation. As such, every debate about the meaning of sustainability reminds us that we must continuously evaluate and revaluate the way we approach modern life, as individuals, as members of our communities, and as global citizens. It also provides a clear call to combat nefarious applications (or misapplications) of sustainability not only because they are misguided in terms of their social, economic, environmental, and political aims and consequences but because they create, expand, or enforce epistemological fault lines. "Sustainable golf," in this sense, may be objectionable as a misinterpretation of sustainability (used as an external, objective measure) but also because it presents an internal, untenable contradiction.

Second, recognizing that the meaning of sustainability is contingent and open to varying interpretations suggests that instead of debating the "right" definition for sustainability, we would do better to explore the multitude meanings of sustainability as they emerge *in practice*. If we approach sustainability as a form of language game (Wittgenstein 2001) or as a concept in the specific sense intended by Deleuze and Guattari (1994)—something that is "created or thought anew in relation to every particular event, insight, experience or problem" (Stagoll 2005, p. 50)— we are required to shift the weight of analysis from evaluating different meanings through the lens of definitional fidelity to tracing the ways in which such meanings emerge through the actions and experiences of different actors. Such experiences, it has been argued on these pages, are increasingly mediated or refracted by interactive technologies. In what follows, I outline those refracted meanings while pointing to their implications. My analysis foregrounds the ways in which the different meanings address complexity, futurity, and agency—the central features of *any* meaning of sustainability. Table 6.1 summarizes the main conclusions of the analysis.

Table 6.1 The main features of the four meanings of sustainability as they emerge from sustainability's mediation by interactive media

Domain	Behavior	Reason	Experience	Imagination
Type of interactions	Persuasive	Synoptic	Resonant	Worldmaking
Desired impact	Behavior change	Informed solutions	Emotional investment	Ontological agency
The user is characterized as...	A creature of habit	A rational agent	An embodied subject	An imaginative shaper
The meaning of sustainability is...	A restored balance between humans and nonhumans	A complex, "wicked" problem	A felt embeddedness	An imaginary
Signature design features	Prompt, nudge, tunneling	Spatiotemporal manipulation based on scientific defensibility	Presence, immersion, diagetic consistency	Ambiguity, open-endedness, unfinishedness

FIRST REFRACTION: SUSTAINABILITY AS RESTORED BALANCE

Chapter 2 discussed interactive media that aim to impact user behavior. What were referred to in the chapter as persuasive interactions, target the "attitude-behavior gap" (Kollmuss and Agyeman 2002): the ostensible gap between people's awareness of, and willingness to act on environmental issues and their lack of actual action. Media designers working from this perspective often elect to intervene on the "action" side of the gap, prompting or nudging users to act in sustainable ways. The examples presented in the chapter illustrate an escalating range of behavioral interventions. What they have in common is a view of users as less-than-rational agents, unwilling or incapable of changing their behavior. Users, in other words, are perceived as creatures of habit, locked into unsustainable behavior patterns and in need of a behavioral jolt. The task of designers is to provide that jolt and help steer users in the right, pro-environmental direction.

Interactive media such as Coralog or greenMeter provide users with just-in-time information about the environmental impacts of their choices with the expectation that this would lead users to take appropriate action. Technologies such as Flower Lamp, StepGreen, or BinCam aim to persuade users to adopt or give up certain activities by deploying physical or social nudges. And finally, technologies such as Erratic Radio attempt to determine behavior outcomes by severely limiting what the user can do, or by "tunneling" (Fogg 2003) users toward particular, predetermined outcomes. Taken to their logical end, persuasive interactions emerge as "technologies of behavior" (Skinner 1971) that seek to reliably promote particular outcomes in the firm belief that the outcomes are justified and agreeable, that is, reasonable and valid according to scientifically articulated contingencies. What is considered sustainable, in this context, is largely derived from scientific calculuses of material processes, inputs, and outcomes and described in quantitative terms: tones, liters, particles per, and so forth.

Refracted by persuasive interactions, sustainability appears as the restoring of the delicate balance between the material needs of humans and nonhumans. To be sure, this image was already implied by the Brundtland Commission's definition of sustainability, and seems to rest on a more traditional ecological view of nature as a system tending toward equilibrium. This planetary equilibrium, the "sacred balance" in Canadian environmentalist David Suzuki's words (2007), was acutely disturbed by the practices and outcomes of industrial modernity—by the irresponsible over-extraction of material resources, the fetishization of economic growth at all costs, the externalization of environmental costs from economic models and financial calculations, and, most significantly in this context, the thoughtless wastefulness that accompanies consumer culture. The unraveling "web of life," however, may still be sutured and the disrupted balance may be restored if modern society learns to rein in its distasteful appetite for conspicuous consumption. For those designing persuasive interactions, this may be achieved by greening and optimizing the cumulative outcome of multiple individual choices across a host of everyday domains.

As discussed in Chap. 2, despite evidence that consumptive behaviors are conditioned by interacting social, economic, and political systems, persuasive interactions rarely target those larger systems as a whole. Instead, they simplify the complexity of sustainability into "bite-sized" behavioral modifications. This has clear advantages: acting on the prompts given by persuasive interactions provides users with a tangible sense of achievement. Making conscious decisions to reduce the consumption of electricity, discard less garbage, recycle more effectively, or take public transit instead of driving a car, provides tangible financial, ecological, and moral rewards. Successful sustainable actions may even increase an individual's perceived self-efficacy and thus help mitigate or even counteract the dreaded "magnitudinal gap" that was discussed in Chap. 2. Doing *something* is surely better than doing *nothing*. On the other hand, oversimplifying sustainability may impart a false sense of achievement while distracting individuals from the more urgent collective task: the decarbonization of the global economy as a whole. This is, of course, a familiar gripe that could be aimed at almost every sustainable micro-action (and feed right back into the "magnitudinal gap"): "You reduce, reuse and recycle. You turn down plastic and paper. You avoid out-of-season grapes. You do all the right things. Good. Just know that it won't save the tuna, protect the rain forest or stop global warming. The changes necessary are so large and profound that they are beyond the reach of individual action" (Wagner 2011, Sep. 7). Furthermore, given that the majority of greenhouse gas (GHG) emissions are produced by industry ("Just 100 companies responsible for 71% of global emissions," writes *The Guardian*),[2] and that changes to industry often necessitate appropriate legislation and regulation, focusing on micro-actions may effectively conceal the need for deeper interventions: "While we busy ourselves greening our personal lives, fossil fuel corporations are rendering these efforts irrelevant" (Lukacs 2017, July 17). Intervening in individual lifestyle choices, it follows, may yield measurable outcomes, but is unlikely to achieve a rapprochement between economic and ecological rationalities (Gorz 2012, p. 33). Efforts to restore the "sacred balance" may start by raising awareness to the social and environmental consequences of modern life, and nudge individuals to make more sustainable choices, but sustainability does not end with individuals acting alone (Brulle 2010; Dourish 2010; Hulme 2009).

SECOND REFRACTION: SUSTAINABILITY AS A COMPLEX PROBLEM

In contrast with the interactive media discussed in Chap. 2, those discussed in Chap. 3 aim to communicate the complexity of sustainability in its fullest. Instead of simplifying sustainability into small achievable actions, the synoptic interactions discussed in the chapter zoom out to reveal a comprehensive, holistic image of sustainability's multiple integrated dimensions. Buckminster Fuller's *World Game* was not designed as a "technology of behavior," and neither were *SimCity* and MetroQuest. Their mission is more pedagogical, electing to teach and demonstrate the intricacies of sustainability instead of motivating users to take particular pro-environmental actions. Through the synoptic interactions offered by

[2] Riley (2017, July 10).

such games and simulations, sustainability appears as a complex, layered, and multidimensional problem that can be solved, or at least managed, by rational agents using the tools and techniques of science. In this view, sustainability involves a host of human and nonhuman actors, linear and nonlinear phenomena, emergent interdependencies, and significant uncertainty. To fully engage with sustainability, one has to first understand it, or at least gain a better sense of its contours and constitutive elements. This is why this particular meaning of sustainability is almost always communicated with spatial and temporal elements: to understand sustainability comprehensively, users are asked to grasp its expansive nature, how it connects seemingly unrelated systems, and how it does so across various spatial and temporal domains (see also Tomlinson 2010).

The communicative emphasis on the interrelated spatiotemporal elements that constitute sustainability as a complex problem reveals the reliance of this particular meaning of sustainability on scientific data and models. As a balance to restore (the first refracted meaning), sustainability relies on science to provide it with unambiguous criteria for success, outlining the ways in which the "sacred balance" may be restored. These criteria, in turn, inform the goals of persuasive interactions. In contrast, as a complex problem, sustainability relies on scientific knowledge to first outline and then fill in the simulated problem space. Scientific fidelity, or what was referred to in Chap. 4 as scientific defensibility (Sheppard 2005), is used to anchor the procedural enactment of sustainability—how it can be performed as a rule-based spatiotemporal dilemma. While relying on scientific knowledge, synoptic interactions seem much more open to alternative types of knowledge than persuasive interactions. There is more room for ambiguity and uncertainty in the solutions offered by synoptic interactions than in those offered by persuasive interactions. This is evident in the way the two types of interactions articulate the future: persuasive interactions tend to mobilize a singular view of a sustainable future—the future that will be brought about by following the behavioral script. Synoptic interactions, in contrast, indicate a multiplicity of sustainable futures which are made available to users for prodding, exploring, and pursuing (Bendor 2018a). Persuasive interactions unfold like an arrow; synoptic interactions fan out.

The fact that as a complex problem sustainability lends itself to multiple perspectives and possible "solutions" has important implications in terms of user agency. The *World Game* developed by Buckminster Fuller, for instance, is global in scope, but it offers much less granularity than *SimCity*.

The latter may present a relatively smaller space to manipulate yet affords players with more possibilities (or "affordances"). MetroQuest presents an even narrower slice of urbanity, but does so in a way that explicitly supports policymaking since it pertains to an actual existing place (and not a mere "everycity") and makes use of available policy levers. The narrower the focus of synoptic interactions, the more concrete the agentic modalities they offer. With that said, the agency afforded by synoptic interactions may not be as immediate as that offered by persuasive interactions. No amount of playing *SimCity* will reduce one's environmental footprint; in fact, in terms of energy use, it may achieve the opposite. However, to the player of *World Game*, *SimCity*, or MetroQuest, agency unfolds across multiple spatiotemporal regimes, and since it is embedded in numerous domains, it offers a much more comprehensive view of potential intervention strategies. Players of *SimCity*, in other words, may not reduce their environmental footprint by playing the game, but they may gain a more astute understanding of how different policy levers may be used to promote sustainability. Such a notion of agency presupposes users as rational agents, able to conjure a mental model of sustainability and act on it in a systemic manner.

THIRD REFRACTION: SUSTAINABILITY AS FELT EMBEDDEDNESS

While persuasive interactions seek to promote sustainable behaviors, and synoptic interactions seek to build mental models that capture and convey the emergent complexity of sustainability, the resonant interactions discussed in Chap. 4 aim to create an experience of sustainability (or of unsustainability). To those flying over a posthuman Paris, facing a forest fire from the perspective of a tree, or peering at the aftermath of a virtual oil spill, the issues entailed by sustainability *feel* real. The price of emotional resonance, however, is often paid for by a diminished adherence to complexity as visceral impact takes precedence over cerebral nuance. Yet, while resonant interactions may not communicate the emergent complexity of sustainability as a (more or less) comprehensive set of manipulable parameters, they do make complexity tangible through the presence, availability, actions, and responses of on-screen entities. These entities choreograph situations that first trigger emotional responses and then chart "mattering maps" (Grossberg 1992) that direct resonance into particular objects. As sustainability becomes less and less abstract—less of a statistical phenomenon and more of a tangible state—it emerges as the embodied impression of inhabiting a space alongside other entities. Sustainability, in other words, appears as a form of felt embeddedness.

Since the notion of sustainability as felt embeddedness centers on the human body as "a means to communicate with the world" (Merleau-Ponty 1962, p. 106), resonant interactions treat the user as an embodied subject whose perception and knowledge of the world are built of tacit, felt relations. For this reason, resonant interactions make use of a range of sensorial triggers and visual, audial, and tactile cues to locate the user within consistent, coherent, believable, and evocative situations. These are designed to help users gain new perspectives about the world. They "thicken" both the perceiving subject and the perceived object, and as result, the relations that unfold in and through sustainability become personally meaningful and salient. No longer mere proposition, sustainability becomes a felt relation, but one that does not promote action over judgment (as is the case with persuasive interactions), nor perception over sensation (as is the case with synoptic interactions). With the development of "smoother" virtual multisensorial experiences, and with the integration of more captivating narrative frameworks and on-screen activities, we can imagine that the appeal and impact of resonant interactions will only grow.

An important aspect of resonant interactions is the way they materialize temporality. Peering into a VR viewfinder becomes a form of time traveling. Users are transported to different times and places, and allowed to experience and enact different spatiotemporal configurations. With each such configuration, a new set of relations is made tangible, and with them a different dimension of sustainability comes to light. In this sense, although the media discussed in Chap. 4 may not fully embody what media theorist Friedrich Kittler calls "time axis manipulation" (Kittler 2017; see also Krämer 2006), they do evoke a sense of temporal fluidity—the sense that our being is not limited to a single place or time. So while the synoptic interactions discussed in Chap. 3 convey a sense of temporality by allowing users to manipulate systems over extensive periods of (virtual) time, the resonant interactions discussed in Chap. 4 pour a measure of felt relationality into the movement of time. Temporality, as consequence, appears more like Bergsonian "duration" than physical time (Bergson 2007; see also Canales 2015).

From the perspective of resonant interactions, to be embedded in a situation, in contrast with observing or contemplating it from the outside, may produce what Stuart Candy (2010, p. 75) calls "felt insight." This is an important rationale for using resonant interactions to augment the more cerebral, infocentric communicative processes that characterize existing public engagement with sustainability. MetroQuest, the sustainability backcasting tool discussed in Chap. 3, for instance, allows users to switch between two visual perspectives: a bird's eye overview of an urban block (suggesting synoptic interactions) and a street-level view (consistent with resonant interactions). While the latter invites users to engage with the issues more personally, the former counteracts potentials for NIMBYism—at least theoretically. The success of the Owl in engaging residents of San Francisco on the proposed developments of Market Street is further testimony to the value of resonant interactions as means to draw the public's attention to the policy question by making it feel "real." Such invitations are even more evocative when resonant interactions are applied with dramatic flair. An oil spill on one's favorite beach may only be simulated, but it certainly feels concrete, significant, and urgent.

FOURTH REFRACTION: SUSTAINABILITY AS AN IMAGINARY

The media discussed in Chap. 5 are experimental in nature. They seek to break new ground in terms of both the experiences they offer and the image of sustainability they conjure. Both ambitions, however, are premised in the value of the imagination in achieving escape velocity or plotting what Deleuze and Guattari (1987) call "lines of flight." The imagination, in other words, is understood as a platform for transformation, key to envisioning and pursuing alternative social, economic, cultural, and political configurations. Sustainability, in turn, appears as an organizing principle for such alternative configurations—as an imaginary: a framework of values, beliefs, and worldviews that give meaning to social practices (Taylor 2004). As such, sustainability provides us with an alternative foundational narrative for organizing everyday life and the institutions charged with maintaining it. The content of such institutions may vary in relation to the particular circumstances in which they are embedded, but the principles that tie them together into a coherent meshwork are oriented by underlying notions of social origins, values, priorities, and norms. Such notions are often produced by the kind of stories we tell ourselves and with which we come to understand our collective identity and goals. It is in this sense that we can understand, for instance, Ehrenfeld's (2008) suggestion that sustainability be understood as "*the possibility that humans and other life will flourish on the Earth forever*" (p. 53; emphasis in origin). In Ehrenfeld's view, sustainability opens up a civilizational narrative according to which the possibility to flourish (and not at the expense of others' ability to do the same) serves as the underlying, organizing metaphor for contemporary society. As the cardinal element of an alternative imaginary, the possibility to flourish serves to calibrate society's moral and ethical compass.

As was emphasized in Chap. 5, the sustainability imaginary is inherently plural: there could be more than one version of sustainability. Each version may derive from different worldviews and may coexist peacefully or in competition with the other ones. The media discussed in Chap. 5 seek to make this plurality visible by criticizing existing imaginaries or offering new ones as evidence of the dynamic, socially constructed nature of sustainability. This kind of pluralization has important ontological implications since how we understand sustainability is premised in how we understand the world more generally. A plurality of beliefs about the very nature of the world foreshadows a plurality of sustainabilities. No less important, and given that sustainability is implicated in efforts to manage and (re)distribute resources, this ontological plurality has political and ethical implications: how we see the world, and how we understand sustainability are intimately related, and shape the ways in which we envision and pursue ways to act. Worldmaking interactions aim to make these connections visible, legible, and therefore malleable. And it is the latter, or more precisely, the possibility of the latter that carries the agentic modalities implicit in this particular meaning of sustainability. Worldmaking interactions may potentiate new forms of "ontological agency" (Maggs 2014; Vervoort et al. 2015).

Although worldmaking interactions foreground the extent to which sustainability is implicated in the social construction of reality (Berger and Luckmann 1966/1989; Couldry and Hepp 2017), they do not immediately lend themselves to programmatic action (and should therefore not be evaluated based on their capacity to do so). The path from the kind of experiences offered by worldmaking interactions to the kind of practices required to bring about a deep transformation of our social imaginaries and the world they disclose is not singular nor straightforward. Exploring future domestic technologies or speculative organisms, opening multiple doors onto the future, or even debating different future possibilities for a quasi-fictional town, may not translate into the kind of immediate action desired by persuasive interactions. But the capacity to recognize the deep contingency of the world and to imagine that it could be otherwise is the seedbed of social transformation. It may not be sufficient but certainly is necessary for opening up new ways to envision and ultimately pursue change. In this sense, the unfinishedness of media deploying worldmaking interactions—the way they make use of open-endedness, ambiguity, and multiplicity to trigger the imagination—serves as a homology both to the nature of our imaginaries and to the nature of the world.

UNCERTAINTY AND HOPE AFTER THE INFORMATION DEFICIT MODEL

The point of departure for this book was that technology and sustainability are entangled. By looking closer into the design and use of interactive media for sustainability, we are now better positioned to see just how the two are entangled. On the one hand, different notions of sustainability influence processes of design, shaping what Feenberg (2017) calls the design code of technologies: "the stabilized intersection of social choice with technical specification" (pp. 73–74). Whether they make their views explicit or not, designers subscribe to particular understandings of sustainability, and these inform their design goals, processes, and outcomes. On the other hand, different designs mediate, shift, or refract the meaning of sustainability for their users. Through their mediated interactions, users interpret and perform new understandings of sustainability. What emerges from this circular and continuous process of mediatization (Couldry and Hepp 2017) is an image of sustainability as a multiplicity: a discursive field in which various actors, operating in different domains, derive different understandings of sustainability which they mobilize in pursuit of different goals. No single meaning of sustainability is more correct than the others, just as none of the design strategies illustrated above can singlehandedly bring about the kind of societal transformations we so desperately need. The four design strategies complement each other, and in some cases could even be combined. As briefly discussed above, MetroQuest, the sustainability decision-support tool, deploys both synoptic and resonant interactions. *Fort McMoney*, the interactive documentary game, includes elements associated with synoptic, resonant, and worldmaking interactions, and so does *Sustainability in an Imaginary World*, the interactive installation. The question, then, is not which type of interactions is more effective but which is more appropriate in light of the goals established by media designers and other relevant stakeholders, and as derived from the way they interpret sustainability. If designers believe that the environmental crises we face are predominantly an outcome of consumer culture, then persuasive interventions in the lifestyle choices we make may yield meaningful results. If designers believe that what society needs is to better understand the potential change levers operating in and through interacting social, economic, ecological, and political systems, then synoptic interactions seem promising. If designers believe that effective societal action is lagging for a lack of felt urgency, then resonant interactions may unlock and unleash strong motivational forces. And if designers believe that sustainability is held back by the lingering dominance of a fundamental set of cultural values and beliefs, then seeding ontological agency with world-making interactions offers an intriguing path to explore.

To even consider the appropriateness of different design interventions against the backdrop of sustainability's semiotic pluralism provides further evidence that the information deficit model is indeed on its way out. There is simply no justification to pursue the kind of one-size-fits-all solutions the model undersigns. But the model's demise also signals an end to the kind of certainty it exudes. As described in Chap. 2, the behavioral cascade that lies at the heart of the information deficit model (see Image 2.1) projected an unwavering faith in linear causality: if only interventions followed the model's logic and sequence, positive results were certain to follow. Once the inevitability dissipates, however, designed interventions must adapt, and as consequence, both media users and media designers must accept a significant measure of uncertainty (Bendor 2018b).

Although some persuasive interactions seek to inform users about the environmental consequences of their behavior, such designs do so while replacing the universal, one-size-fits-all communication strategy associated with the information deficit model with more nuanced, customizable information delivery mechanisms. Even so, persuasive design's coupling of information delivery with clear micro-behavioral instructions does not guarantee that these instructions will translate into concrete action (indeed, as discussed in Chap. 2, this is quite often the case). As long as the user is granted a significant measure of autonomy, fantasies about a reliable "technology of sustainable behavior" will remain just that. While persuasive interactions reluctantly admit a measure of uncertainty into their behavioral interventions, synoptic interactions embrace uncertainty by design. This is largely derived from the way synoptic interactions reject the information deficit model's reductive simplification of sustainability, electing instead to simulate and materialize multiple possibilities that mirror the considerable variability of interacting human and nonhuman systems. As the degree of variability grows, the certainty of success implicit in any single path of action declines. Whereas synoptic interactions retain, to some extent, the logocentrism of the information deficit model, resonant interactions reject it altogether. They appeal to emotion instead of reason, seeking to promote action by pushing embodied, affective buttons instead of merely informing users about the consequences of their behavior. But even successful stimulations of felt urgency do not immediately translate to urgent action. Here, too, certainty gives way to mere possibility. We find a similar pattern in worldmaking interactions. By targeting the imaginaries that organize social, cultural, economic, and political institutions instead of the institutions themselves, worldmaking interactions are much more abstract than the other types of interaction. But by assuming that different imaginaries are ontologically equivalent, and by communicating the possibility that multiple imaginaries can coexist, worldmaking interactions defy the kind of deterministic singularity of action that floated the information deficit model. In this sense, worldmaking interactions embody the essence of what appeared in *Sustainability in an Imaginary World* as the credo of the "literary" world: "no world is carved in stone."

By rejecting the universality, reductiveness, logocentrism, and determinism of the information deficit model, the contemporary design and use of interactive media for sustainability appear to embrace uncertainty, thus mirroring the semiotic pluralism of sustainability itself. This does, however, raise important questions. On the one hand, granting "equal rights" to different meanings of sustainability and the practical strategies they unfold may feed into a crippling hesitation to act in the face of impending civilizational collapse. How can we act if we are uncertain about whether our actions are indeed sustainable (i.e., whether they adhere to the "right" version of sustainability)? And why should we maintain space for uncertainty when evidence of the scope and impending consequences of climate change is so overwhelming? For those impacted by extreme weather events, floods, raging forest fires, or drought—indeed for all those who share the planet—time for decisive action is running out. Can we afford to spend precious time and energy reflecting on the appropriateness of this or that meaning of sustainability—to prioritize catching up our thinking with our living when the very possibility of living on this planet is in jeopardy? On the other hand, sustainability's strength as a concept derives precisely from its capacity to nourish a multiplicity of meanings and paths of action; its staying power and gravity, as Ulrich Grober (2012, p. 195) writes, is premised in its flexibility: "The word contains everything that matters." Is it not wise, then, to draw on sustainability's semiotic pluralism to expand and renew our repertoire of designerly (and political) interventions? Should we not hedge against the failure of any single path of action and build up, in the process, our social learning and resilience?

Answers to these questions remain beyond the scope of the present work, but I have little doubt that they will influence the design of the next generation of interactive media for sustainability. We can already see such questions raised in the context of transition design, transformative deign, civic media design, and other cognate fields. Nonetheless, perhaps the real issue is not whether we should embrace or reject uncertainty wholesale— to some extent, we have no choice; the world is ripe with uncertainty—but whether we can identify domains in which uncertainty is not an impediment but a requisite for transformative action. This is the gist of author Rebecca Solnit's observation that "Hope locates itself in the premises that we don't know what will happen and that in the spaciousness of uncertainty is room to act" (2016, p. xiv). Uncertainty, in other words, bears a fundamental freedom to shape reality, a sense that things can be otherwise, that even the most immutable centers of power will eventually buckle under the pressure of collective action. No matter the precise interpretation of sustainability held by designers of interactive media, we cannot do without such hope.

Bibliography

Bendor, R. (2018a). Interaction Design for Sustainability Futures: Towards Worldmaking Interactions. In M. Hazas & L. P. Nathan (Eds.), *Digital Technology and Sustainability: Engaging the Paradox* (pp. 205–216). New York: Routledge.

Bendor, R. (2018b). Sustainability, Hope and Designerly Action in the Anthropocene. *Interactions*, 25(3), 82–84.

Berger, P. L., & Luckmann, T. (1966/1989). *The Social Construction of Reality: A Treatise in the Sociology of Knowledge*. New York: Anchor Books.

Bergson, H. (2007). *The Creative Mind: An Introduction to Metaphysics* (trans: Andison, M. L.). Mineola: Dover Publications.

Brulle, R. J. (2010). From Environmental Campaigns to Advancing the Public Dialog: Environmental Communication for Civic Engagement. *Environmental Communication*, 4(1), 82–98.

Callon, M., Lascoumes, P., & Barthe, Y. (2009). *Acting in an Uncertain World: An Essay on Technical Democracy* (trans: Burchell, G.). Cambridge, MA: MIT Press.

Canales, J. (2015). *The Physicist and the Philosopher: Einstein, Bergson, and the Debate That Changed Our Understanding of Time.* Princeton/Oxford: Princeton University Press.

Candy, S. (2010). *The Futures of Everyday Life: Politics and the Design of Experiential Scenarios.* PhD dissertation submitted at the University of Hawaii, Manoa.

Caradonna, J. L. (2014). *Sustainability: A History.* New York: Oxford University Press.

Connelly, S. (2007). Mapping Sustainable Development as a Contested Concept. *Local Environment: The International Journal of Justice and Sustainability, 12*(3), 259–278.

Couldry, N., & Hepp, A. (2017). *The Mediated Construction of Reality.* Cambridge/Malden: Polity Press.

Daly, H. E. (1993). Sustainable Growth: An Impossibility Theorem. In H. E. Daly & K. N. Townsend (Eds.), *Valuing the Earth: Economics, Ecology, Ethics* (pp. 267–273). Cambridge, MA: MIT Press.

Deleuze, G., & Guattari, F. (1987). *A Thousand Plateaus: Capitalism and Schizophrenia* (trans: Massumi, B.). Minneapolis: University of Minnesota Press.

Deleuze, G., & Guattari, F. (1994). *What Is Philosophy?* (trans: Tomlinson, H., & Burchell, G.). New York: Columbia University Press.

Dourish, P. (2010). HCI and Environmental Sustainability: The Politics of Design and the Design of Politics. In O. W. Bartelsen & P. Krogh (Eds.), *Proceedings of DIS 2010* (pp. 1–10). New York: ACM.

Dryzek, J. S. (2013). *The Politics of the Earth: Environmental Discourses* (3rd ed.). Oxford/New York: Oxford University Press.

Ehrenfeld, J. (2008). *Sustainability by Design: A Subversive Strategy for Transforming Our Consumer Culture.* New Haven: Yale University Press.

Feenberg, A. (2017). *Technosystem: The Social Life of Reason.* Cambridge, MA: Harvard University Press.

Fogg, B. J. (2003). *Persuasive Technology: Using Computers to Change What We Think and Do.* Amsterdam/Boston: Morgan Kaufmann Publishers.

Funtowicz, S. O., & Ravetz, J. R. (1993). Science for the Post-Normal Age. *Futures, 25*(7), 739–755.

Gibbons, M., Limoges, C., Nowotny, H., Schwartzman, S., Scott, P., & Trow, M. (1994). *The New Production of Knowledge: The Dynamics of Science and Research in Contemporary Societies.* London: Sage.

Gorz, A. (2012). *Capitalism, Socialism, Ecology* (trans: Chalmers, M.). London/New York: Verso.

Grober, U. (2012). *Sustainability: A Cultural History* (trans: Cunningham, R.). Totnes: Green Books.

Grossberg, L. (1992). *We Gotta Get Out of This Place: Popular Conservatism and Postmodern Culture*. New York: Routledge.

Hacking, I. (1999). *The Social Construction of What?* Cambridge, MA: Harvard University Press.

Hulme, M. (2009). *Why We Disagree About Climate Change: Understanding Controversy, Inaction and Opportunity*. Cambridge/New York: Cambridge University Press.

Jacobs, M. (1999). Sustainable Development as a Contested Concept. In A. Dobson (Ed.), *Fairness and Futurity: Essays on Environmental Sustainability and Social Justice* (pp. 21–45). Oxford/New York: Oxford University Press.

Kittler, F. (2017). Real Time Analysis, Time Axis Manipulation. *Cultural Politics, 13*(1), 1–18.

Kollmuss, A., & Agyeman, J. (2002). Mind the Gap: Why Do People Act Environmentally and What Are the Barriers to Pro-environmental Behavior? *Environmental Education Research, 8*(3), 239–260.

Krämer, S. (2006). The Cultural Techniques of Time Axis Manipulation: On Friedrich Kittler's Conception of Media. *Theory, Culture & Society, 23*(7–8), 93–109.

Lukacs, M. (2017, July 17). Neoliberalism Has Conned Us into Fighting Climate Change as Individuals. *The Guardian*. Retrieved from https://www.theguardian.com/environment/true-north/2017/jul/17/neoliberalism-has-conned-us-into-fighting-climate-change-as-individuals

Maggs, D. (2014). *Artists of the Floating World*. PhD dissertation submitted at the University of British Columbia, Vancouver.

Maggs, D., & Robinson, J. (2016). Recalibrating the Anthropocene: Sustainability in an Imaginary World. *Environmental Philosophy, 13*(2), 175–194.

Marshall, J. D., & Toffel, M. W. (2005). Framing the Elusive Concept of Sustainability: A Sustainability Hierarchy. *Environmental Science and Technology, 39*(3), 673–682.

Merleau-Ponty, M. (1962). *Phenomenology of Perception* (trans: Smith, C.). London/Henley: Routledge/Kegan Paul.

Miller, T. R. (2013). Constructing Sustainability Science: Emerging Perspectives and Research Trajectories. *Sustainability Science, 8*(2), 279–293.

Mittler, D. (2001). Hijacking Sustainability? Planners and the Promise and Failure of Local Agenda 21. In A. Layard, S. Davoudi, & S. Batty (Eds.), *Planning for a Sustainable Future* (pp. 53–60). London/New York: Taylor & Francis.

Nardi, B., Tomlinson, B., Patterson, D., Chen, J., Pargman, D., Raghavan, B., & Penzenstadler, B. (forthcoming). Computing Within Limits. *Communications of the ACM*.

Parr, A. (2009). *Hijacking Sustainability*. Cambridge, MA: MIT Press.

Riley, T. (2017, July 10). Just 100 Companies Responsible for 71% of Global Emissions, Study Says. *The Guardian*. Retrieved from https://www.

theguardian.com/sustainable-business/2017/jul/10/100-fossil-fuel-companies-investors-responsible-71-global-emissions-cdp-study-climate-change

Robinson, J. (2004). Squaring the Circle? Some Thoughts on the Idea of Sustainable Development. *Ecological Economics, 48*, 369–384.

Rockström, J., & Klum, M. (2015). *Big World, Small Planet: Abundance Within Planetary Boundaries*. New Haven: Yale University Press.

Rockström, J., Steffen, W., Noone, K., Persson, Å., Chapin, F. S., III, Lambin, E. F., et al. (2009). A Safe Operating Space for Humanity. *Nature, 461*, 472–475.

Sheppard, S. R. J. (2005). Landscape Visualisation and Climate Change: The Potential for Influencing Perceptions and Behaviour. *Environmental Science and Policy, 8*, 637–654.

Skinner, B. F. (1971). *Beyond Freedom and Dignity*. Indianapolis/Cambridge: Hackett.

Solnit, R. (2016). *Hope in the Dark: Untold Histories, Wild Possibilities* (3rd ed.). Chicago: Haymarket Books.

Stagoll, C. (2005). Concepts. In A. Parr (Ed.), *The Deleuze Dictionary* (pp. 50–51). Edinburgh: Edinburgh University Press.

Suzuki, D. T. (2007). *The Sacred Balance: Rediscovering Our Place in Nature* (3rd ed.). Vancouver: David Suzuki Foundation/Greystone Books.

Taylor, C. (2004). *Modern Social Imaginaries*. Durham: Duke University Press.

Tomlinson, B. (2010). *Greening Through IT: Information Technology for Environmental Sustainability*. Cambridge, MA: MIT Press.

UN. (2015). *Transforming Our World: The 2030 Agenda for Sustainable Development*. Retrieved from https://sustainabledevelopment.un.org/content/documents/21252030%20Agenda%20for%20Sustainable%20Development%20web.pdf

UNWCED. (1987). *Our Common Future: Report of the World Commission on Environment and Development*. New York: Oxford University Press.

Vervoort, J. M., Bendor, R., Kelliher, A., Strik, O., & Helfgott, A. E. R. (2015). Scenarios and the Art of Worldmaking. *Futures, 74*, 62–70.

Wagner, G. (2011, September 7). Going Green But Getting Nowhere. *New York Times*. Retrieved from http://www.nytimes.com/2011/09/08/opinion/going-green-but-getting-nowhere.html

Wittgenstein, L. (2001). *Philosophical Investigations* (trans: Anscombe, G. E. M., 3rd ed.). Oxford/Malden: Blackwell.

BIBLIOGRAPHY

Aarseth, E. J. (1997). *Cybertext: Perspectives on Ergodic Literature*. Baltimore: Johns Hopkins University Press.

Adams, P. C. (1998). Teaching and Learning with SimCity 2000. *Journal of Geography, 97*(2), 47–55.

Agamben, G. (1993). *Infancy and History: The Destruction of Experience* (trans: Heron, L.). London/New York: Verso.

Ajzen, I. (1991). The Theory of Planned Behavior. *Organizational Behavior and Human Decision Processes, 50,* 179–211.

Akrich, M. (1992). The Description of Technical Objects. In W. E. Bijker & J. Law (Eds.), *Shaping Technology/Building Society: Studies in Sociotechnical Change* (pp. 205–224). Cambridge, MA: MIT Press.

Albert, B. (2016, November 14). Eagle Flight Review. *IGN*. Retrieved from http://www.ign.com/articles/2016/11/14/eagle-flight-review

Als, H. (2017, February 13 & 20). Capturing James Baldwin's Legacy Onscreen. *The New Yorker*. Retrieved from http://www.newyorker.com/magazine/2017/02/13/capturing-james-baldwins-legacy-onscreen

Alvarez, A. C. (2017, April 26). Envisioning New Futures: Steve Lambert and Stephen Duncombe on Artistic Activism. *Creative Capital Blog*. Retrieved from http://blog.creative-capital.org/2017/04/envisionng-new-futures-steve-lambert-stephen-duncombe-artistic-activism/

Anderson, A. G. (2014). *Media, Environment and the Network Society*. Basingstoke/New York: Palgrave.

Andrejevic, M. (2016). The Pacification of Interactivity. In D. Barney, G. Coleman, C. Ross, J. Sterne, & T. Tembeck (Eds.), *The Participatory Condition in the*

Digital Age (pp. 187–206). Minneapolis/London: University of Minnesota Press.

Arias-Maldonado, M. (2013). Rethinking Sustainability in the Anthropocene. *Environmental Politics, 22*(3), 428–446.

Arthos, J. (2000). 'To Be Alive When Something Happens': Retrieving Dilthey's Erlebnis. *Janus Head, 3*(1). Retrieved from http://www.janushead.org/3-1/jarthos.cfm

Atanasova, D., & Koteyko, N. (2017). Metaphors in Guardian Online and Mail Online Opinion-Page Content on Climate Change: War, Religion, and Politics. *Environmental Communication, 11*(4), 452–469.

Augé, M. (2015). *The Future.* London/New York: Verso.

Backlund, S., Gyllenswärd, M., Gustafsson, A., Ilstedt Hjelm, S., Mazé, R., & Redström, R. (2006). *STATIC! The Aesthetics of Energy in Everyday Things.* Paper Presented at the Design Research Society International Conference, Lisbon.

Bakhtin, M. M. (1981). *The Dialogic Imagination: Four Essays* (trans: Holquist, M.). Austin: University of Texas Press.

Ball, D. P. (2014, November 12). Burrard Inlet Binoculars Imagine Oil-Slicked Disaster. *The Tyee.* Retrieved from https://thetyee.ca/News/2014/11/12/Burrard-Inlet-Installation/

Bamberg, S., & Möser, G. (2007). Twenty Years After Hines, Hungerford, and Tomera: A New Meta-Analysis of Psycho-social Determinants of Pro-environmental Behaviour. *Journal of Environmental Psychology, 27*, 14–25.

Bandura, A. (1994). Self-Efficacy. In V. S. Ramachaudran (Ed.), *Encyclopedia of Human Behavior* (Vol. 4, pp. 71–81). New York: Academic.

Barreto, M., Karapanos, E., & Nunes, N. (2013). *Why Don't Families Get Along with Eco-feedback Technologies?: A Longitudinal Inquiry, Proceeding of CHItaly '13 (Article No. 16).* New York: ACM.

Barry, A. (2001). *Political Machines: Governing a Technological Society.* New Brunswick: Athlone Press.

Bendor, R. (2012). Analytic and Deictic Approaches to the Design of Sustainability Decision-Making Tools. In *Proceedings of iConference '12* (pp. 215–222). Toronto.

Bendor, R. (2017). Interactive World Disclosure (or, an Interface Is Not a Hammer). In T. Markham & S. Rodgers (Eds.), *Conditions of Mediation: Phenomenological Perspectives on Media* (pp. 211–221). New York: Peter Lang.

Bendor, R. (2018a). Interaction Design for Sustainability Futures: Towards Worldmaking Interactions. In M. Hazas & L. P. Nathan (Eds.), *Digital Technology and Sustainability: Engaging the Paradox* (pp. 205–216). New York: Routledge.

Bendor, R. (2018b). Sustainability, Hope and Designerly Action in the Anthropocene. *Interactions, 25*(3), 82–84.

Bendor, R., Anacleto, J., Facey, D., Fels, S., Herron, T., Maggs, D., et al. (2015). Sustainability in an Imaginary World. *Interactions, 22*(5), 54–57.

Bendor, R., Maggs, D., Peake, R., Robinson, J., & Williams, S. (2017). The Imaginary Worlds of Sustainability: Observations from an Interactive Art Installation. *Ecology and Society, 22*(2), 17.

Benhabib, S. (Ed.). (1996). *Democracy and Difference: Contesting the Boundaries of the Political*. Princeton: Princeton University Press.

Benjamin, W. (1929/1999). The Great Art of Making Things Seem Closer Together. In M. W. Jennings, H. Eiland, & G. Smith (Eds.), *Selected Writings (Vol. 2 Pt. 1)* (p. 248). Cambridge, MA: Belknap Press.

Berdichevsky, D., & Neuenschwander, E. (1999). Toward an Ethics of Persuasive Technology. *Communications of the ACM, 42*(5), 51–58.

Bereitschaft, B. (2016). Gods of the City? Reflecting on City Building Games as an Early Introduction to Urban Systems. *Journal of Geography, 115*(2), 51–60.

Berger, P. L., & Luckmann, T. (1966/1989). *The Social Construction of Reality: A Treatise in the Sociology of Knowledge*. New York: Anchor Books.

Bergin, J. (2000, July). *Fourteen Pedagogical Patterns*. Retrieved from http://csis.pace.edu/~bergin/PedPat1.3.html

Bergson, H. (2007). *The Creative Mind: An Introduction to Metaphysics* (trans: Andison, M. L.). Mineola: Dover Publications.

Bhamra, T., Lilley, D., & Tang, T. (2011). Design for Sustainable Behaviour: Using Products to Change Consumer Behaviour. *The Design Journal, 14*(4), 427–445.

Bickman, L. (1972). Environmental Attitudes and Actions. *The Journal of Social Psychology, 87*(2), 323–324.

Blevis, E. (2007). Sustainable Interaction Design: Invention & Disposal, Renewal & Reuse. In M. B. Rosson & D. Gilmore (Eds.), *Proceedings of CHI 2007* (pp. 503–512). New York: ACM Press.

Blevis, E., Lim, Y.-k., Roedl, D., & Stolterman, E. (2007). Using Design Critique as Research to Link Sustainability and Interactive Technologies. In D. Schuler (Ed.), *Online Communities and Social Computing* (pp. 22–31). Berlin/Heidelberg: Springer.

Bogost, I. (2007). *Persuasive Games: The Expressive Power of Videogames*. Cambridge, MA: MIT Press.

Boks, C., Lilley, D., & Pettersen, I. N. (2015). The Future of Design for Sustainable Behaviour, Revisited. Paper Presented at the 9th EcoDesign International Symposium on Environmentally Conscious Design and Inverse Manufacturing, Tokyo.

Bolter, J. D., & Grusin, R. A. (1999). *Remediation: Understanding New Media*. Cambridge, MA: MIT Press.

Bonneuil, C., & Fressoz, J.-B. (2017). *The Shock of the Anthropocene*. London/New York: Verso.

Bord, R. J., O'Connor, R. E., & Fisher, A. (2000). In What Sense Does the Public Need to Understand Global Climate Change? *Public Understanding of Science, 9*, 205–218.

Borges, J. L. (1999). *Collected Fictions* (trans: Hurley, A.). New York: Penguin.

Bottici, C. (2014). *Imaginal Politics: Images Beyond Imagination and the Imaginary*. New York: Columbia University Press.

Brey, P. (2000). Disclosive Computer Ethics. *Computers and Society, 30*(4), 10–16.

Broomell, S. B., Budescu, D. V., & Por, H.-H. (2015). Personal Experience with Climate Change Predicts Intentions to Act. *Global Environmental Change, 32*, 67–73.

Brulle, R. J. (2010). From Environmental Campaigns to Advancing the Public Dialog: Environmental Communication for Civic Engagement. *Environmental Communication, 4*(1), 82–98.

Brynjarsdóttir, H., Håkansson, M., Pierce, J., Baumer, E. P. S., DiSalvo, C., & Sengers, P. (2012). Sustainably Unpersuaded: How Persuasion Narrows Our Vision of Sustainability. In *Proceedings of CHI '12* (pp. 947–956). New York: ACM.

Bucchi, M. (2008). Of Deficits, Deviations and Dialogues: Theories of Public Communication of Science. In M. Bucchi & B. Trench (Eds.), *Handbook of Public Communication of Science and Technology* (pp. 57–76). London/New York: Routledge.

Buchanan, R. (1985). Declaration by Design: Rhetoric, Argument, and Demonstration in Design Practice. *Design Issues, 2*(1), 4–22.

Bucher, T. (2012). Want to Be on the Top? Algorithmic Power and the Threat of Invisibility on Facebook. *New Media & Society, 14*(7), 1164–1180.

Burdea, G., & Coiffet, P. (2003). *Virtual Reality Technology* (2nd ed.). Hoboken: J. Wiley-Interscience.

Burgess, J., Harrison, C., & Filius, P. (1998). Environmental Communication and the Cultural Politics of Environmental Citizenship. *Environment and Planning A, 30*, 1445–1460.

Büscher, B. (2014). Nature 2.0: Exploring and Theorizing the Links Between New Media and Nature Conservation. *New Media & Society, 18*(5), 726–743.

Busse, D., Mann, S., Nathan, L. P., & Preist, C. (2013). Changing Perspectives on Sustainability: Healthy Debate or Divisive Factions. In *CHI 2013 Extended Abstracts on Human Factors in Computing Systems* (pp. 2505–2508). New York: ACM.

Caesar, M. (1999). *Umberto Eco: Philosophy, Semiotics and the Work of Fiction*. Oxford/Malden, MA: Polity Press.

Callon, M., Lascoumes, P., & Barthe, Y. (2009). *Acting in an Uncertain World: An Essay on Technical Democracy* (trans: Burchell, G.). Cambridge, MA: MIT Press.

Canales, J. (2015). *The Physicist and the Philosopher: Einstein, Bergson, and the Debate That Changed Our Understanding of Time*. Princeton/Oxford: Princeton University Press.

Candy, S. (2010). *The Futures of Everyday Life: Politics and the Design of Experiential Scenarios*. PhD Dissertation Submitted at the University of Hawaii, Manoa.

Cantwell Smith, B. (2002). The Foundations of Computing. In M. Scheutz (Ed.), *Computationalism: New Directions* (pp. 23–58). Cambridge, MA: MIT press.

Capra, F. (1985). Criteria of Systems Thinking. *Futures, 17*(5), 475–478.

Caprara, G. V., Vecchione, M., Capanna, C., & Mebane, M. (2009). Perceived Political Self-Efficacy: Theory, Assessment, and Applications. *European Journal of Social Psychology, 39*(6), 1002–1020.

Caradonna, J. L. (2014). *Sustainability: A History*. New York: Oxford University Press.

Carmichael, J., Tansey, J., & Robinson, J. (2004). An Integrated Assessment Modeling Tool. *Global Environmental Change, 14*, 171–183.

Casey, E. S. (1976). *Imagining: A Phenomenological Study*. Bloomington: Indiana University Press.

Castoriadis, C. (1997). *The Imaginary Institution of Society* (trans: Blamey, K.). Cambridge, MA: MIT Press.

Chapman, D. A., Lickel, B., & Markowitz, E. M. (2017). Reassessing Emotion in Climate Change Communication. *Nature Climate Change, 7*, 850–852.

Chawla, L. (1998). Significant Life Experiences Revisited: A Review of Research on Sources of Environmental Sensitivity. *Environmental Education Research, 4*(4), 369–382.

Chawla, L. (2006). Learning to Love the Natural World Enough to Protect It. *Barn, 2*, 57–78.

Clear, A., Preist, C., Joshi, S., Nathan, L. P., Mann, S., & Nardi, B. A. (2015). Expanding the Boundaries: A SIGCHI HCI & Sustainability Workshop, CHI 2015 Extended Abstracts on Human Factors in Computing Systems (pp. 2373–2376). New York: ACM.

Coeckelbergh, M. (2017). *Using Words and Things: Language and Philosophy of Technology*. London/New York: Routledge.

Connelly, S. (2007). Mapping Sustainable Development as a Contested Concept. *Local Environment: The International Journal of Justice and Sustainability, 12*(3), 259–278.

Connolly, W. E. (2006). Experience & Experiment. *Daedalus, 135*(3), 67–75.

Constine, J. (2017, January 20). Sundance Merges VR with Real Life Through Props, AR, and Vibrating Suits. *TechCrunch*. Retrieved from https://techcrunch.com/2017/01/20/sundance-new-frontier/

Cooper, A. (2004). *The Inmates Are Running the Asylum*. Indianapolis: Sams.

Costikyan, G. (2013). *Uncertainty in Games*. Cambridge, MA: MIT Press.

Couldry, N. (2008). Mediatization or Mediation? Alternative Understandings of the Emergent Space of Digital Storytelling. *New Media & Society, 10*(3), 373–391.

Couldry, N., & Hepp, A. (2017). *The Mediated Construction of Reality*. Cambridge/Malden: Polity Press.

Cox, J. R., & Pezzullo, P. C. (2015). *Environmental Communication and the Public Sphere* (4th ed.). Thousand Oaks: Sage.

Cresswell, T. (2015). *Place: An Introduction* (2nd ed.). Chichester/Malden: Wiley Blackwell.

Crompton, T. (2010). *Common Cause: The Case for Working with Our Cultural Values*. Surrey: WWF-UK.

Crompton, T., & Kasser, T. (2009). *Meeting Environmental Challenges: The Role of Human Identity*. Surrey: WWF-UK.

Cross, K., Gunster, S., Piotrowski, M., & Daub, S. (2015). *News Media and Climate Politics: Civic Engagement and Political Efficacy in a Climate of Reluctant Cynicism*. Vancouver, BC: Canadian Centre for Policy Alternatives. Retrieved from https://www.policyalternatives.ca/sites/default/files/uploads/publications/BC%20Office/2015/09/CCPA-BCNews_Media_Climate_Politics.pdf

Daly, H. E. (1993). Sustainable Growth: An Impossibility Theorem. In H. E. Daly & K. N. Townsend (Eds.), *Valuing the Earth: Economics, Ecology, Ethics* (pp. 267–273). Cambridge, MA: MIT Press.

Damasio, A. R. (1994). *Descartes' Error: Emotion, Reason, and the Human Brain*. New York: Putnam.

Damasio, A. R. (2003). *Looking for Spinoza: Joy, Sorrow, and the Feeling Brain* (1st ed.). Orlando: Harcourt.

Daniels, S. E., & Walker, G. B. (1996). Collaborative Learning: Improving Public Deliberation in Ecosystem-Based Management. *Environmental Impact Assessment Review, 16*(2), 71–102.

Darbellay, F. (2015). Rethinking Inter- and Transdisciplinarity: Undisciplined Knowledge and the Emergence of a New Thought Style. *Futures, 65*, 163–174.

Davidson, R. J. (2000). Cognitive Neuroscience Needs Affective Neuroscience (and Vice Versa). *Brain and Cognition, 42*(1), 89–92.

Davis, J. (2009). Design Methods for Ethical Persuasive Computing. In *Proceedings of the 4th International Conference on Persuasive Technology* (pp. 6–13). New York: ACM.

Dawson, M. (2003). *The Consumer Trap: Big Business Marketing in American Life*. Urbana: University of Illinois Press.

de Oliveira, R., & Carrascal, J. P. (2014). Towards Effective Ethical Behavior Design. In *Proceedings of CHI 2014* (pp. 2149–2154). New York: ACM.

de Rosnay, J. (2011). Symbionomic Evolution: From Complexity and Systems Theory, to Chaos Theory and Coevolution. *World Futures, 67*(4–5), 304–315.

Debord, G. (1994). *The Society of the Spectacle* (trans: Nicholson-Smith, D.). New York: Zone Books.

Deleuze, G., & Guattari, F. (1987). *A Thousand Plateaus: Capitalism and Schizophrenia* (trans: Massumi, B.). Minneapolis: University of Minnesota Press.

Deleuze, G., & Guattari, F. (1994). *What Is Philosophy?* (trans: Tomlinson, H., & Burchell, G.). New York: Columbia University Press.

Descartes, R. (2017). *Meditations on First Philosophy with Selections from the Objections and Replies* (trans: Cottingham, J., 2nd ed.). Cambridge: Cambridge University Press.

Devisch, O. (2008). Should Planners Start Playing Computer Games? Arguments from SimCity and Second Life. *Planning Theory & Practice, 9*(2), 209–226.

Dewey, J. (1934). *Art as Experience.* New York: Paragon.

Dijksterhuis, E. J. (1961). *The Mechanization of the World Picture* (trans: Dikshoorn, C.). London/New York: Oxford University Press.

DiSalvo, C., Sengers, P., & Brynjarsdóttir, H. (2010). Mapping the Landscape of Sustainable HCI. In *Proceedings of CHI 2010* (pp. 1975–1984). New York: ACM Press.

Douglas, M., & Wildavsky, A. B. (1982). *Risk and Culture: An Essay on the Selection of Technical and Environmental Dangers.* Berkeley: University of California Press.

Dourish, P. (2001). *Where the Action Is: The Foundations of Embodied Interaction.* Cambridge, MA: MIT Press.

Dourish, P. (2010). HCI and Environmental Sustainability: The Politics of Design and the Design of Politics. In O. W. Bartelsen & P. Krogh (Eds.), *Proceedings of DIS 2010* (pp. 1–10). New York: ACM.

Downes, E. J., & McMillan, S. (2000). Defining Interactivity: A Qualitative Identification of Key Dimensions. *New Media & Society, 2*(2), 157–179.

Dreyfus, H. L. (1992). *What Computers Still Can't Do: A Critique of Artificial Reason.* Cambridge, MA: MIT Press.

Dryzek, J. S. (2013). *The Politics of the Earth: Environmental Discourses* (3rd ed.). Oxford/New York: Oxford University Press.

Dunne, A., & Raby, F. (2013). *Speculative Everything: Design, Fiction, and Social Dreaming.* Cambridge, MA: MIT Press.

Dyball, R., & Newell, B. (2015). *Understanding Human Ecology: A Systems Approach to Sustainability.* London/New York: Routledge.

Eco, U. (1989). *The Open Work* (trans: Cancogni, A.). Cambridge, MA: Harvard University Press.

Eco, U. (1994). *Six Walks in the Fictional Woods.* Cambridge, MA: Harvard University Press.

Edwards, P. N. (2010). *A Vast Machine: Computer Models, Climate Data, and the Politics of Global Warming.* Cambridge, MA: MIT Press.

Egan, P. J., & Mullin, M. (2012). Turning Personal Experience into Political Attitudes: The Effect of Local Weather on Americans' Perceptions About Global Warming. *The Journal of Politics, 74*(3), 796–809.

Egan, P. J., & Mullin, M. (2017). Climate Change: US Public Opinion. *Annual Review of Political Science, 20*, 209–227.

Eglash, R., Croissant, J., Di Chiro, G., & Fouché, R. (Eds.). (2004). *Appropriating Technology: Vernacular Science and Social Power.* Minneapolis: University of Minnesota Press.

Ehgartner, U., Gould, P., & Hudson, M. (2017). On the Obsolescence of Human Beings in Sustainable Development. *Global Discourse, 7*(1), 66–83.

Ehrenfeld, J. (2008). *Sustainability by Design: A Subversive Strategy for Transforming Our Consumer Culture.* New Haven: Yale University Press.

Eisenstein, C. (2011). *Sacred Economics: Money, Gift, and Society in the Age of Transition.* Berkeley: North Atlantic Books.

Erikson, J. (2012, March 5). Overcoming Barriers to a Green Economy. *SustainAbility.* Retrieved from http://sustainability.com/ourwork/insights/overcoming-barriers-to-a-green-economy/

Ermi, L., & Mäyrä, F. (2005). Fundamental Components of the Gameplay Experience: Analysing Immersion. Paper Presented in the 2005 DiGRA Conference: Changing Views – World in Play, Vancouver.

Espinosa, A., & Walker, J. (2011). *A Complexity Approach to Sustainability: Theory and Application.* London: Imperial College Press.

Ewen, S. (2001). *Captains of Consciousness: Advertising and the Social Roots of the Consumer Culture* (25th anniversary ed.). New York: Basic Books.

Farley, H. M., & Smith, Z. A. (2014). *Sustainability: If It's Everything, Is It Nothing?* London/New York: Routledge.

Feenberg, A. (1999). *Questioning Technology.* London/New York: Routledge.

Feenberg, A. (2002). *Transforming Technology: A Critical Theory Revisited.* New York: Oxford University Press.

Feenberg, A. (2009). Peter-Paul Verbeek: Review of *What Things Do. Human Studies, 32*(2), 225–228.

Feenberg, A. (2010). Between Reason and Experience. In *Between Reason and Experience: Essays in Technology and Modernity* (pp. 181–218). Cambridge, MA: MIT.

Feenberg, A. (2017). *Technosystem: The Social Life of Reason.* Cambridge, MA: Harvard University Press.

Figueres, C., Schellnhuber, H. J., Whiteman, G., Rockström, J., Hobley, A., & Rahmstorf, S. (2017). Three Years to Safeguard Our Climate. *Nature, 546,* 593–595.

Fletcher, R. (2017). Gaming Conservation: Nature 2.0 Confronts Nature-Deficit Disorder. *Geoforum, 79,* 153–162.

Flew, T., & Smith, R. (2011). *New Media: An Introduction (Canadian Edition)* (2nd ed.). Oxford/New York: Oxford University Press.

Fogg, B. J. (2003). *Persuasive Technology: Using Computers to Change What We Think and Do.* Amsterdam/Boston: Morgan Kaufmann Publishers.

Ford, L. (2016). "Unlimiting the Bounds": The Panorama and the Balloon View. *The Public Domain Review.* Retrieved from https://publicdomainreview.org/2016/08/03/unlimiting-the-bounds-the-panorama-and-the-balloon-view

Forrester, J. W. (1998). Designing the Future. Lecture delivered December 15, 1998, at Universidad de Sevilla, Sevilla.

Foster, D., Lawson, S., Blythe, M., & Cairns, P. (2010). Wattsup?: Motivating Reductions in Domestic Energy Consumption Using Social Networks. In *Proceedings of NordiCHI 2010* (pp. 178–187). New York: ACM.

Freud, S. (1920). *A General Introduction to Psychoanalysis* (trans: Hall, G. S.). New York: Boni & Liveright Publishers.

Frick, T. (2016). *Designing for Sustainability: A Guide to Building Greener Digital Products and Services*. Sebastopol: O'Reilly.

Friedberg, A. (2006). *The Virtual Window: from Alberti to Microsoft*. Cambridge, MA: MIT Press.

Friedman, B. (1996). Value-Sensitive Design. *Interactions, 3*(6), 17–23.

Friedman, B., & Nissenbaum, H. (1996). Bias in Computer Systems. *ACM Transactions on Information Systems, 14*(3), 330–347.

Fritsch, J., & Brynskov, M. (2011). Between Experience, Affect, and Information: Experimental Urban Interfaces in the Climate Change Debate. In M. Foth, L. Forlano, C. Satchell, & M. Gibbs (Eds.), *From Social Butterfly to Engaged Citizen: Urban Informatics, Social Media, Ubiquitous Computing, and Mobile Technology to Support Citizen Engagement* (pp. 115–134). Cambridge, MA: MIT Press.

Froehlich, J., Findlater, L., & Landay, J. (2010). The Design of Eco-feedback Technology. In *Proceedings of CHI 2010* (pp. 1999–2008). New York: ACM.

Fuad-Luke, A. (2009). *Design Activism: Beautiful Strangeness for a Sustainable World*. London/Sterling: Earthscan.

Fuller, R. B. (1969). *50 Years of the Design Science Revolution and the World Game: A Collection of Articles and Papers on Design*. Carbondale: World Resources Inventory, Southern Illinois University.

Fuller, R. B. (1999). *Your Private Sky* (edited by Joachim Krausse & Claude Lichtenstein). Zurich: Lars Muller Publishers.

Funk, J. (2017). Assessing Public Forecasts to Encourage Accountability: The Case of MIT's Technology Review. *PLoS One, 12*(8), e0183038.

Funtowicz, S. O., & Ravetz, J. R. (1993). Science for the Post-Normal Age. *Futures, 25*(7), 739–755.

Gaber, J. (2007). Simulating Planning – SimCity as a Pedagogical Tool. *Journal of Planning Education and Research, 27*(2), 113–121.

Gabrielli, S., & Maimone, R. (2014). Designing a Context-Aware Mobile Application for Eco-driving. In *Proceedings of ICCASA 14* (pp. 102–104). Brussles: ICST.

Gadamer, H. G. (2004). *Truth and Method* (trans: Weinsheimer, J., & Marshall, D. G., 3rd ed.). New York/London: Continuum.

Galloway, A. R. (2012). *The Interface Effect*. Malden/Cambridge: Polity.

Gamberini, L., Spagnolli, A., Corradi, N., Jacucci, G., Tusa, G., Mikkola, T., et al. (2012). Tailoring Feedback to Users' Actions in a Persuasive Game for Household Electricity Conservation. In M. Bang & E. L. Ragnemalm (Eds.), *Persuasive Technology. Design for Health and Safety. PERSUASIVE 2012* (pp. 100–111). Berlin/Heidelberg: Springer.

Ganglbauer, E., Reitberger, W., & Fitzpatrick, G. (2013). An Activist Lens for Sustainability: From Changing Individuals to Changing the Environment. In S. Berkovsky & J. Freyne (Eds.), *PERSUASIVE 2013* (pp. 63–68). Berlin/Heidelberg: Springer.

Garrard, G., Handwerk, G., & Wilke, S. (2014). Introduction: "Imagining Anew: Challenges of Representing the Anthropocene". *Environmental Humanities, 5*, 149–153.

Gauntlett, D. (2005). *Moving Experiences: Media Effects and Beyond* (2nd ed.). Eastleigh/Bloomington: John Libbey Pub.

Geelen, D., Keyson, D., Boess, S., & Brezet, H. (2012). Exploring the Use of a Game to Stimulate Energy Saving in Households. *Journal of Design Research, 10*(1–2), 102–120.

Gershon, I., & Bell, J. A. (2013). Introduction: The Newness of New Media. *Culture, Theory and Critique, 54*(3), 259–264.

Ghosh, A. (2016, October 28). Where Is the Fiction About Climate Change? *The Guardian*. Retrieved from https://www.theguardian.com/books/2016/oct/28/amitav-ghosh-where-is-the-fiction-about-climate-change-_-__jmp0_

Gibbons, M., Limoges, C., Nowotny, H., Schwartzman, S., Scott, P., & Trow, M. (1994). *The New Production of Knowledge: The Dynamics of Science and Research in Contemporary Societies*. London: Sage.

Gitelman, L. (2006). *Always Already New: Media, History and the Data of Culture*. Cambridge, MA: MIT Press.

Goldberg, H. (2013, November 26). Where Film Marries Video Game. *New York Times*. Retrieved from http://www.nytimes.com/2013/11/27/arts/video-games/where-film-marries-video-game.html

Goldstein, J. (1999). Emergence as a Construct: History and Issues. *Emergence, 1*(1), 49–72.

Goodman, N. (1978). *Ways of Worldmaking*. Indianapolis: Hackett Pub. Co.

Gordon, E., Schirra, S., & Hollander, J. (2011). Immersive Planning: A Conceptual Model for Designing Public Participation with New Technologies. *Environment and Planning B: Planning and Design, 38*, 505–519.

Gorz, A. (2012). *Capitalism, Socialism, Ecology* (trans: Chalmers, M.). London/New York: Verso.

Gregory, J., & Miller, S. (1998). *Science in Public: Communication, Culture, and Credibility*. New York: Perseus.

Grober, U. (2012). *Sustainability: A Cultural History* (trans: Cunningham, R.). Totnes: Green Books.

Grossberg, L. (1992). *We Gotta Get Out of This Place: Popular Conservatism and Postmodern Culture*. New York: Routledge.

Guerin, D. A., Yust, B. L., & Coopet, J. G. (2000). Occupant Predictors of Household Energy Behavior and Consumption Change as Found in Energy Studies Since 1975. *Family and Consumer Sciences Research Journal, 29*(1), 48–80.

Haas Lyons, S., Walsh, M., Aleman, E., & Robinson, J. (2014). Exploring Regional Futures: Lessons from Metropolitan Chicago's Online MetroQuest. *Technological Forecasting and Social Change, 82*, 23–33.

Habermas, J. (1990). *Moral Consciousness and Communicative Action* (trans: Lenhardt, C., & Nicholsen, S. W.). Cambridge, MA: MIT Press.

Habermas, J. (1996). Three Normative Models of Democracy. In S. Benhabib (Ed.), *Democracy and Difference: Contesting the Boundaries of the Political* (pp. 21–30). Princeton: Princeton University Press.

Habermas, J. (2001). Truth and Society: The Discursive Redemption of Factual Claims to Validity. In *On the Pragmatics of Social Interaction: Preliminary Studies in the Theory of Communicative Action* (pp. 85–103). Cambridge, MA: MIT Press.

Hacking, I. (1999). *The Social Construction of What?* Cambridge, MA: Harvard University Press.

Haider, J. (2016). The Shaping of Environmental Information in Social Media: Affordances and Technologies of Self-Control. *Environmental Communication, 10*(4), 473–491.

Haiven, M. (2014). *Crises of Imagination, Crises of Power: Capitalism, Creativity and the Commons*. London: Zed Books.

Haiven, M., & Khasnabish, A. (2014). *The Radical Imagination: Social Movement Research in the Age of Austerity*. London: Zed Books.

Hamari, J., Koivisto, J., & Pakkanen, T. (2014). Do Persuasive Technologies Persuade? – A Review of Empirical Studies. In A. Spagnolli, L. Chittaro, & L. Gamberini (Eds.), *Persuasive Technology. PERSUASIVE 2014* (pp. 137–142). Cham: Springer.

Hansen, M. B. N. (2004). *New Philosophy for New Media*. Cambridge, MA: MIT Press.

Hansen, A. (2010). *Environment, Media and Communication*. London/New York: Routledge.

Hansen, A., & Cox, J. R. (Eds.). (2015). *The Routledge Handbook of Environment and Communication*. London/New York: Routledge.

Harman, G. (2007). *Heidegger Explained: From Phenomenon to Thing*. Chicago: Open Court.

Hart, P. S., & Leiserowitz, A. A. (2009). Finding the Teachable Moment: An Analysis of Information-Seeking Behavior on Global Warming Related Websites

During the Release of the Day After Tomorrow. *Environmental Communication: A Journal of Nature and Culture, 3*(3), 355–366.

Hassenzahl, M., & Laschke, M. (2015). Pleasurable Troublemakers. In S. P. Walz & S. Deterding (Eds.), *The Gameful World: Approaches, Issues, Applications* (pp. 167–195). London/Cambridge, MA: MIT Press.

Hazas, M., & Nathan, L. P. (2018a). Introduction: Digital Technology and Sustainability: Engaging the Paradox. In M. Hazas & L. P. Nathan (Eds.), *Digital Technology and Sustainability: Engaging the Paradox* (pp. 3–13). New York: Routledge.

Hazas, M., & Nathan, L. P. (Eds.). (2018b). *Digital Technology and Sustainability: Engaging the Paradox.* New York: Routledge.

Hazas, M., Bernheim Brush, A. J., & Scott, J. (2012). Sustainability Does Not Begin with the Individual. *Interactions, 19*(5), 14–17.

Heidegger, M. (1962). *Being and Time* (trans: Macquarrie, J., & Robinson, E.). San Francisco: HarperSanFrancisco.

Heidegger, M. (1971a). Building, Dwelling, Thinking. In *Poetry, Language, Thought* (trans: Hofstadter, A., pp. 145–161). New York: Harper and Row.

Heidegger, M. (1971b). The Thing. In *Poetry, Language, Thought* (trans: Hofstadter, A., pp. 165–182). New York: Harper and Row.

Heidegger, M. (1977). *The Question Concerning Technology, and Other Essays* (trans: Lovitt, W., 1st ed.). New York: Harper & Row.

Herrnstein Smith, B. (2015, May 6). What Was "Close Reading"? A Century of Method in Literary Studies. Paper Presented at the Heyman Center, Columbia University, New York, for a Digital Humanities Workshop series "On Method".3.

Hilty, L. M., & Aebischer, B. (Eds.). (2015). *ICT Innovations for Sustainability.* Cham: Springer.

Hiraoka, T., Terakado, Y., Matsumoto, S., & Yamabe, S. (2009). Quantitative Evaluation of Eco-driving on Fuel Consumption Based on Driving Simulator Experiments. In *Proceedings of the 16th World Congress on Intelligent Transport Systems* (pp. 21–25). Washington, DC: ITS.

Hirschkop, K. (2004). Justice and Drama: On Bakhtin as a Complement to Habermas. In N. Crossley & J. M. Roberts (Eds.), *After Habermas: New Perspectives on the Public Sphere* (pp. 49–66). Oxford/Malden: Blackwell Publishing/The Sociological Review.

Hjarvard, S. (2013). *The Mediatization of Culture and Society.* Abington/New York: Routledge.

Hornsey, M. J., Harris, E. A., Bain, P. G., & Fielding, K. S. (2016). Meta-Analyses of the Determinants and Outcomes of Belief in Climate Change. *Nature Climate Change, 6,* 622.

Hulme, M. (2009). *Why We Disagree About Climate Change: Understanding Controversy, Inaction and Opportunity.* Cambridge/New York: Cambridge University Press.

Ihde, D. (1990). *Technology and the Lifeworld: From Garden to Earth*. Bloomington: Indiana University Press.

Ingold, T. (2000). *The Perception of the Environment: Essays on Livelihood, Dwelling and Skill*. London/New York: Routledge.

Innes, J. E., & Booher, D. E. (2010). *Planning with Complexity: An Introduction to Collaborative Rationality for Public Policy*. London /New York: Routledge.

IPCC. (2014). *Climate Change 2014 Mitigation of Climate Change: Working Group III Contribution to the Fifth Assessment Report of the Intergovernmental Panel on Climate Change*. New York: Cambridge University Press.

IPCC. (2015). *Climate Change 2014: Synthesis Report. Contribution of Working Groups I, II and III to the Fifth Assessment Report of the Intergovernmental Panel on Climate Change*. Retrieved from Geneva, https://www.ipcc.ch/pdf/assessment-report/ar5/syr/SYR_AR5_FINAL_full_wcover.pdf

Issa, T., Isaias, P., & Issa, T. (Eds.). (2017). *Sustainability, Green IT and Education Strategies in the Twenty-First Century*. Cham: Springer.

Jacobs, M. (1999). Sustainable Development as a Contested Concept. In A. Dobson (Ed.), *Fairness and Futurity: Essays on Environmental Sustainability and Social Justice* (pp. 21–45). Oxford/New York: Oxford University Press.

Jaffe, Z. (2017, February 23). Tree VR "Grows" with SUBPAC at Sundance. Retrieved from http://subpac.com/tree-vr-grows-subpacsundance/

Jang, S. M., & Hart, P. S. (2015). Polarized Frames on "Climate Change" and "Global Warming" Across Countries and States: Evidence from Twitter Big Data. *Global Environmental Change, 32*, 11–17.

Jauss, H. R. (1982). *Toward an Aesthetic of Reception* (trans: Bahti, T.). Minneapolis: University of Minnesota Press.

Jay, M. (2005). *Songs of Experience: Modern American and European Variations on a Universal Theme*. Berkeley: University of California Press.

Jégou, F., & Gouache, C. (2015). Envisioning as an Enabling Tool for Social Empowerment and Sustainable Democracy. In V. W. Thoresen, R. J. Didham, J. Klein, & D. Doyle (Eds.), *Responsible Living: Concepts, Education and Future Perspectives* (pp. 253–271). Cham: Springer.

Jensen, J. F. (2008). *The Concept of Interactivity – Revisited: Four New Typologies for a New Media Landscape*. Paper Presented at uxTV 2008, Silicon Valley.

Johnson, S. (2001). *Emergence: The Connected Lives of Ants, Brains, Cities, and Software*. New York: Scribner.

Joshi, K. (2017, June 7). Caring About Climate Change: It's Time to Build a Bridge Between Data and Emotion. *The Guardian*. Retrieved from https://www.theguardian.com/commentisfree/2017/jun/07/caring-about-climate-change-its-time-to-build-a-bridge-between-data-and-emotion

Kahan, D. M., Jenkins-Smith, H., & Braman, D. (2011). Cultural Cognition of Scientific Consensus. *Journal of Risk Research, 14*(2), 147–174.

Kahneman, D. (2011). *Thinking, Fast and Slow*. New York: Farrar, Straus and Giroux.

Karppinen, P., & Oinas-Kukkonen, H. (2013). Three Approaches to Ethical Considerations in the Design of Behavior Change Support Systems. In S. Berkovsky & J. Freyne (Eds.), *PERSUASIVE 2013* (pp. 87–98). Berlin/Heidelberg: Springer.

Katz-Kimchi, M., & Manosevitch, I. (2015). Mobilizing Facebook Users Against Facebook's Energy Policy: The Case of Greenpeace Unfriend Coal Campaign. *Environmental Communication, 9*(2), 248–267.

Kearney, R. (1998). *Poetics of Imagining: Modern to Post-Modern* (2nd ed.). New York: Fordham University Press.

Kelly, K. (1995, June). Gossip Is Philosophy. Kevin Kelly Talks to Brian Eno. *Wired, 3.05,* 146–151, 204–149.

Kelly, M. R. (2012). Climate Change Communication Research Here and Now: A Reflection on Where We Came From and Where We Are Going. *Applied Environmental Education & Communication, 11*(3–4), 117–118.

Kim, P. (1992). Does Advertising Work: A Review of the Evidence. *Journal of Consumer Marketing, 9*(4), 5–21.

Kim, M., & Shin, J. (2016). The Pedagogical Benefits of SimCity in Urban Geography Education. *Journal of Geography, 115*(2), 39–50.

Kim, T., Hong, H., & Magerko, B. (2010). Design Requirements for Ambient Display That Supports Sustainable Lifestyle. In *Proceedings of DIS 2010* (pp. 103–112). New York: ACM.

Kirilenko, A. P., & Stepchenkova, S. O. (2014). Public Microblogging on Climate Change: One Year of Twitter Worldwide. *Global Environmental Change, 26,* 171–182.

Kitchin, R., & Dodge, M. (2011). *Code/Space: Software and Everyday Life.* Cambridge, MA: MIT Press.

Kitchin, R., & Freundschuh, S. (2000). *Cognitive Mapping: Past, Present and Future.* London: Routledge.

Kittler, F. (2017). Real Time Analysis, Time Axis Manipulation. *Cultural Politics, 13*(1), 1–18.

Klein, N. (2014). *This Changes Everything: Capitalism vs. the Climate.* Toronto: Alfred A. Knopf.

Knowles, B., Blair, L., Coulton, P., & Lochrie, M. (2014). Rethinking Plan A for Sustainable HCI. In *Proceedings of CHI 2014* (pp. 3593–3596). New York: ACM.

Koger, S. M., & Winter, D. D. N. (2010). *The Psychology of Environmental Problems: Psychology for Sustainability* (3rd ed.). New York/London: Psychology Press.

Kollmuss, A., & Agyeman, J. (2002). Mind the Gap: Why Do People Act Environmentally and What Are the Barriers to Pro-environmental Behavior? *Environmental Education Research, 8*(3), 239–260.

Krämer, S. (2006). The Cultural Techniques of Time Axis Manipulation: On Friedrich Kittler's Conception of Media. *Theory, Culture & Society, 23*(7–8), 93–109.

Krimsky, S., & Plough, A. L. (1988). *Environmental Hazards: Communicating Risks as a Social Process.* Dover: Auburn House.

Krueger, M. W. (1977/2003). Responsive Environments. In N. Wardrip-Fruin & N. Montfort (Eds.), *The New Media Reader* (pp. 379–389).Cambridge, MA/ London: MIT Press.

Kuhn, T. S. (1962). *The Structure of Scientific Revolutions.* Chicago: University of Chicago Press.

Kuijer, L., & Bakker, C. (2015). Of Chalk and Cheese: Behaviour Change and Practice Theory in Sustainable Design. *International Journal of Sustainable Engineering, 8*(3), 219–230.

Kumar, K. (1978). *Prophecy and Progress: The Sociology of Industrial and Post-Industrial Society.* Harmondsworth/New York: Penguin.

Kwastek, K. (2008). Interactivity – A Word in Process. In C. Sommerer, L. C. Jain, & L. Mignonneau (Eds.), *The Art and Science of Interface and Interaction Design* (pp. 15–26). Berlin & Heidelberg: Springer.

Lakoff, G. (2004). *Don't Think of an Elephant!: Know Your Values and Frame the Debate: The Essential Guide for Progressives.* White River Junction: Chelsea Green Pub. Co.

Lakoff, G., & Johnson, M. (1999). *Philosophy in the Flesh: The Embodied Mind and Its Challenge to Western Thought.* New York: Basic Books.

Lakoff, G., & Johnson, M. (1980/2003). *Metaphors We Live by.* Chicago: University of Chicago Press.

Latour, B. (1988). *The Pasteurization of France* (trans: Sheridan, A., & Law, J.). Cambridge, MA: Harvard University Press.

Latour, B. (1993). *We Have Never Been Modern.* Cambridge, MA: Harvard University Press.

Latour, B. (2004). *Politics of Nature: How to Bring the Sciences into Democracy* (trans: Porter, C.). Cambridge, MA: Harvard University Press.

Latour, B. (2005). *Reassembling the Social: An Introduction to Actor-Network-Theory.* Oxford/New York: Oxford University Press.

Latour, B. (2014). Agency at the Time of the Anthropocene. *New Literary History, 45*(1), 1–18.

Latour, B. (2017). *Facing Gaia: Eight Lectures on the New Climatic Regime* (trans: Porter, C.). Cambridge/Medford: Polity.

Lauwaert, M. (2007). Challenge Everything? Construction Play in Will Wright's SimCity. *Games and Culture, 20*(3), 194–212.

Lavender, T. (2010, July 18). Digital Media Students Want to Raise Your Carbon Consciousness. *Vancouver Observer.* Retrieved from http://www. vancouverobserver.com/blogs/megabytes/2010/07/18/digital-media-students-want-raise-your-carbon-consciousness

Lawson, P. J., & Flocke, S. A. (2009). Teachable Moments for Health Behavior Change: A Concept Analysis. *Patient Education and Counseling, 76*, 25–30.

Lee, H., Lee, W., & Lim, Y.-K. (2010). The Effect of Eco-driving System Towards Sustainable Driving Behavior. In *Proceedings of CHI 2010* (pp. 4255–4260). New York: ACM.

Leiss, W. (1972). *The Domination of Nature*. New York: G. Braziller.

Lessig, L. (2006). *Code: Version 2.0* (2nd ed.). New York: Basic Books.

Levin, S. (1999). *Fragile Dominion: Complexity and the Commons*. Cambridge, MA: Perseus Books.

Lewis, S. L., & Maslin, M. A. (2015). Defining the Anthropocene. *Nature, 519*(7542), 171–180.

Li, Y., Johnson, E. J., & Zaval, L. (2011). Local Warming: Daily Temperature Change Influences Belief in Global Warming. *Psychological Science, 22*(4), 454–459.

Lilienfeld, R. (1975). Systems Theory as an Ideology. *Social Research, 42*(4), 637–660.

Lilley, D., & Wilson, G. T. (2013). Integrating Ethics into Design for Sustainable Behaviour. *Journal of Design Research, 11*(3), 278–299.

Lilley, D., Lofthouse, V., & Bhamra, T. (2005). Towards Instinctive Sustainable Product Use. Paper Presented at the 2nd International Conference: Sustainability Creating the Culture, Aberdeen.

Lister, M., Dovey, J., Giddings, S., Grant, I., & Kelly, K. (2008). *New media: A Critical Introduction* (2nd ed.). Abingdon/New York: Routledge.

Lockton, D. (2013). *Design with Intent: A Design Pattern Toolkit for Environmental & Social Behaviour Change*. PhD Dissertation Submitted at Brunel University, London.

Lorince, J. (2013, March 6). Emergence (and Some Devastation) in Sim City. *Motivate. Play.* Retrieved from http://www.motivateplay.com/2013/03/emergence-and-some-devastation-in-sim-city

Losh, E. M. (2009). *Virtualpolitik: An Electronic History of Government Media-Making in a Time of War, Scandal, Disaster, Miscommunication, and Mistakes*. Cambridge, MA: MIT Press.

Lövbrand, E., Beck, S., Chilvers, J., Forsyth, T., Hedrén, J., Hulme, M., et al. (2015). Who Speaks for the Future of Earth? How Critical Social Science Can Extend the Conversation on the Anthropocene. *Global Environmental Change, 32*, 211–218.

Löwgren, J. (2009). Toward an Articulation of Interaction Esthetics. *New Review of Hypermedia and Multimedia, 15*(2), 129–146.

Lukacs, M. (2017, July 17). Neoliberalism Has Conned Us into Fighting Climate Change as Individuals. *The Guardian*. Retrieved from https://www.theguardian.com/environment/true-north/2017/jul/17/neoliberalism-has-conned-us-into-fighting-climate-change-as-individuals

Lupia, A., McCubbins, M. D., & Popkin, S. L. (2000). Beyond Rationality: Reason and the Study of Politics. In A. Lupia, M. D. McCubbins, & S. L. Popkin (Eds.), *Elements of Reason: Cognition, Choice, and the Bounds of Rationality* (pp. 1–20). Cambridge/New York: Cambridge University Press.

MacKinnon, J. B. (2013). *The Once and Future World: Nature as It Was, as It Is, as It Could Be.* Boston/New York: Houghton Mifflin Harcourt.

Maggs, D. (2014). *Artists of the Floating World.* PhD dissertation submitted at the University of British Columbia, Vancouver.

Maggs, D., & Robinson, J. (2016). Recalibrating the Anthropocene: Sustainability in an Imaginary World. *Environmental Philosophy, 13*(2), 175–194.

Malm, A. (2016). *Fossil Capital: The Rise of Steam-Power and the Roots of Global Warming.* London/New York: Verso.

Malpas, J. (2006). *Heidegger's Topology: Being, Place, World.* Cambridge, MA: MIT Press.

Malpass, M. (2017). *Critical Design in Context: History, Theory, and Practices.* London/New York: Bloomsbury.

Manjoo, F. (2013, March 4). The New SimCity Is Totally Addictive and Crazily Comprehensive. *Slate.* Retrieved from http://www.slate.com/articles/technology/technology/2013/03/simcity_review_the_new_version_of_the_classic_game_is_totally_addictive.html

Mankoff, J. C., Fussell, S. R., Dillahunt, T., Glaves, R., Grevet, C., Johnson, M., et al. (2010). StepGreen.org: Increasing Energy Saving Behaviors Via Social Networks. In *Proceedings of the Fourth International AAAI Conference on Weblogs and Social Media* (pp. 106–113). Palo Alto: AAAI.

Mann, G., & Wainwright, J. (2018). *Climate Leviathan: A Political Theory of Our Planetary Future.* London/New York: Verso.

Manovich, L. (2001). *The Language of New Media.* Cambridge, MA: MIT Press.

Marcuse, H. (1964/1991). *One-Dimensional Man: Studies in the Ideology of Advanced Industrial Society.* Boston: Beacon Press.

Marcuse, H. (2001). Beyond One-Dimensional Man. In D. Kellner (Ed.), *Towards a Critical Theory of Society* (pp. 111–120). London/New York: Routledge.

Marshall, J. D., & Toffel, M. W. (2005). Framing the Elusive Concept of Sustainability: A Sustainability Hierarchy. *Environmental Science and Technology, 39*(3), 673–682.

Mason, S. (2013). On Games and Links: Extending the Vocabulary of Agency and Immersion in Interactive Narratives. In H. Koenitz, T. I. Sezen, G. Ferri, M. Haahr, D. Sezen, & G. v. Çatak (Eds.), *Interactive Storytelling* (pp. 25–34). Heidelberg: Springer.

Massey, N. (2012, March 12). SimCity 2013 Players Will Face Tough Choices on Energy and Environment. *Scientific American.* Retrieved from https://www.scientificamerican.com/article/simcity-2013-players-face-tough-energy-environment-choices

Maxwell, R. (2013, March 8). Spatial Orientation and the Brain: The Effects of Map Reading and Navigation. *GIS Lounge*. Retrieved from https://www. gislounge.com/spatial-orientation-and-the-brain-the-effects-of-map-reading-and-navigation/

Mayer, I. S. (2008). The Gaming of Policy and the Politics of Gaming: A Review. *Simulation & Gaming, 40*(6), 825–862.

Mazé, R. (2010). *Static! Designing for Energy Awareness*. Stockholm: Arvinius Förlag.

McCarthy, J., & Wright, P. (2004). *Technology as Experience*. Cambridge, MA: MIT Press.

McDermott, J. (2014, February 9). Using the New SimCity, 6 Urban Planners Battle for Bragging Rights. *Co.Design*. Retrieved from https://www.fastcoexist.com/1681515/using-the-new-sim-city-6-urban-planners-battle-for-bragging-rights

McGinn, R. E. (1990). What Is Technology? In L. A. Hickman (Ed.), *Technology as a Human Affair* (pp. 10–25). New York: McGraw Hill.

McKenzie-Mohr, D. (2000). Fostering Sustainable Behavior Through Community-Based Social Marketing. *American Psychologist, 55*(5), 531–537.

McLuhan, M. (1962). *The Gutenberg Galaxy; The Making of Typographic Man*. Toronto: University of Toronto Press.

McLuhan, M. (1964). *Understanding Media; The Extensions of Man*. New York: McGraw-Hill.

McMahan, A. (2003). Immersion, Engagement, and Presence: A Method for Analyzing 3-D Video Games. In M. J. P. Wolf & B. Perron (Eds.), *The Video Game Theory Reader* (pp. 67–86). New York/London: Routledge.

McNichol, T. (2010). The Art Museum as Laboratory for Reimagining a Sustainable Future. In T. Thatchenkery, D. L. Cooperrider, & M. Avital (Eds.), *Positive Design and Appreciative Construction: From Sustainable Development to Sustainable Value* (pp. 177–193). Bingley: Emerald Group Publishing Limited.

McRae, L., Freeman, R., & Marconi, V. (2016). The Living Planet Index. In N. Oerlemans (Ed.), *Living Planet Report 2016: Risk and Resilience in a New Era*. Gland: WWF International.

Meadows, D. H. (2008). *Thinking in Systems: A Primer*. White River Junction: Chelsea Green Publishing.

Meadows, D. H., Meadows, D. L., Randers, J., & Behrens, W. W., III. (1972). *The Limits to Growth; A Report for the Club of Rome's Project on the Predicament of Mankind*. New York: Universe Books.

Merchant, C. (1989). *The Death of Nature: Women, Ecology, and the Scientific Revolution*. New York: Harper & Row.

Merchant, C. (2008). *Ecology* (2nd ed.). Amherst: Humanity Books.

Merleau-Ponty, M. (1962). *Phenomenology of Perception* (trans: Smith, C.). London/Henley: Routledge/Kegan Paul.

Merleau-Ponty, M. (1968). *The Visible and the Invisible* (trans: Lingis, A.). Evanston: Northwestern University Press.

Midgley, G. (2003). *Systems Thinking* (Vol. 4). London/Thousand Oaks: Sage.

Milbank, D. (2016, November 18). Trump's Fake-News Presidency. *The Washington Post.* Retrieved from https://www.washingtonpost.com/opinions/trumps-fake-news-presidency/2016/11/18/72cc7b14-ad96-11e6-977a-1030f822fc35_story.html

Miller, T. R. (2013). Constructing Sustainability Science: Emerging Perspectives and Research Trajectories. *Sustainability Science, 8*(2), 279–293.

Minsky, M. (2006). *The Emotion Machine: Commonsense Thinking, Artificial Intelligence, and the Future of the Human Mind.* New York: Simon & Schuster.

Mittler, D. (2001). Hijacking Sustainability? Planners and the Promise and Failure of Local Agenda 21. In A. Layard, S. Davoudi, & S. Batty (Eds.), *Planning for a Sustainable Future* (pp. 53–60). London/New York: Taylor & Francis.

Monbiot, G. (2017, September 9). How Do We Get Out of This Mess? *The Guardian.* Retrieved from https://www.theguardian.com/books/2017/sep/09/george-monbiot-how-de-we-get-out-of-this-mess

Moores, S. (2012). *Media, Place and Mobility.* Basingstoke/New York: Palgrave Macmillan.

Morozov, E. (2013, March 2). The Perils of Perfection. *New York Times.* Retrieved from http://www.nytimes.com/2013/03/03/opinion/sunday/the-perils-of-perfection.html

Moser, S. C. (2010). Communicating Climate Change: History, Challenges, Process and Future Directions. *Wiley Interdisciplinary Reviews: Climate Change, 1*(1), 31–53.

Moser, S. C. (2016). What More Is There to Say? Reflections on Climate Change Communication Research and Practice in the Second Decade of the 21st Century. *Wiley Interdisciplinary Reviews: Climate Change, 7*(3), 345–369.

Moser, S. C., & Dilling, L. (2007a). Introduction. In S. C. Moser & L. Dilling (Eds.), *Creating a Climate for Change: Communicating Climate Change and Facilitating Social Change* (pp. 1–27). Cambridge/New York: Cambridge University Press.

Moser, S. C., & Dilling, L. (Eds.). (2007b). *Creating a Climate for Change: Communicating Climate Change and Facilitating Social Change.* Cambridge/New York: Cambridge University Press.

Moser, S. C., Daniels, C., Pike, C., & Huva, A. (2016). *Here-Now-Us: Visualizing Sea Level Rise and Adaptation Using the OWL Technology in Marin County,* Santa Cruz. Retrieved from https://climateaccess.org/sites/default/files/Here%20Now%20Us%20Project%20and%20Research%20Summary.pdf

Murray, J. H. (1997). *Hamlet on the Holodeck: The Future of Narrative in Cyberspace.* Cambridge, MA: MIT Press.

Murray, J. H. (2012). *Inventing the Medium: Principles of Interaction Design as a Cultural Practice*. Cambridge, MA: MIT Press.

Myers, T. A., Maibach, E. W., Roser-Renouf, C., Akerlof, K., & Leiserowitz, A. A. (2013). The Relationship Between Personal Experience and Belief in the Reality of Global Warming. *Nature Climate Change, 3,* 343–347.

Nardi, B., & Ekebia, H. (2018). Developing a Political Economy Perspective for Sustainable HCI. In M. Hazas & L. P. Nathan (Eds.), *Digital Technology and Sustainability: Engaging the Paradox* (pp. 86–102). New York: Routledge.

Nardi, B., Tomlinson, B., Patterson, D., Chen, J., Pargman, D., Raghavan, B., & Penzenstadler, B. (forthcoming). Computing Within Limits. *Communications of the ACM.*

Nash, K. (2014). Clicking on the Real: Telling Stories and Engaging Audiences Through Interactive Documentaries. Retrieved from http://eprints.lse.ac.uk/71175/1/blogs.lse.ac.uk-Clicking%20on%20the%20real%20telling%20stories%20and%20engaging%20audiences%20through%20interactive%20documentaries.pdf.

Nelson, A. (2016). The Praxis of Sustainability Citizenship. In R. Horne, J. Fien, B. B. Beza, & A. Nelson (Eds.), *Sustainability Citizenship in Cities: Theory and Practice* (pp. 17–28). London/New York: Routledge.

Newig, J. (2007). Does Public Participation in Environmental Decisions Lead to Improved Environmental Quality? *CCP (Communication, Cooperation, Participation. Research and Practice for a Sustainable Future), 1,* 51–71.

Niedderer, K., Ludden, G., Clune, S. J., Lockton, D., Mackrill, J., Morris, A., et al. (2016). Design for Behaviour Change as a Driver for Sustainable Innovation: Challenges and Opportunities for Implementation in the Private and Public Sectors. *International Journal of Design, 10*(2), 67–85.

Nilsson, E. M., & Jakobsson, A. (2011). Simulated Sustainable Societies: Students' Reflections on Creating Future Cities in Computer Games. *Journal of Science Education and Technology, 20*(1), 33–50.

Nodder, C. (2013). *Evil by Design: Interaction Design to Lead Us into Temptation.* Indianapolis: Wiley.

Nogueira, P. (2015). Ways of Feeling: Audience's Meaning Making in Interactive Documentary Through an Analysis of Fort McMoney. *Punctum, 1*(1), 79–93.

Norman, D. A. (2002). *The Design of Everyday Things* (2nd ed.). New York: Doubleday.

Norman, D. A. (2011). *Living with Complexity.* Cambridge, MA: MIT Press.

Nowotny, H., Scott, P., & Gibbons, M. (2001). *Re-thinking Science: Knowledge and the Public in the Age of Uncertainty.* Cambridge: Polity Press.

Nyhan, B., & Reifler, J. (2010). When Corrections Fail: The Persistence of Political Misperceptions. *Political Behavior, 32*(2), 303–330.

O'Neil, P. (2014, April 30). Kinder Morgan Pipeline Application Says Oil Spills Can Have Both Negative and Positive Effects. *Vancouver Sun.* Retrieved from http://www.vancouversun.com/news/Kinder+Morgan+pipeline+application +says+spills+have+both+negative+positive+effects/9793673/story.html

O'Neill, S., & Boykoff, M. (2011). The Role of New Media in Engaging the Public with Climate Change. In L. Whitmarsh, S. O'Neill, & I. Lorenzoni (Eds.), *Engaging the Public with Climate Change: Behaviour Change and Communication* (pp. 233–251). London/Washington, DC: Earthscan.

Odum, E. P. (1983). *Basic Ecology*. Philadelphia: Saunders College Pub.

Oinas-Kukkonen, H. (2013). A Foundation for the Study of Behavior Change Support Systems. *Personal and Ubiquitous Computing, 17*(6), 1223–1235.

Oinas-Kukkonen, H., & Harjumaa, M. (2009). Persuasive Systems Design: Key Issues, Process Model, and System Features. *Communications of the Association for Information Systems, 24*(article 28), 485–500.

Ortega y Gasset, J. (1941). *Toward a Philosophy of History*. New York: W.W. Norton.

Osborn, A. F. (1993). *Applied Imagination; Principles and Procedures of Creative Problem-Solving* (3rd rev. ed.). Buffalo: Creative Education Foundation.

Otto, E. C., & Wilkinson, A. (2012). Harnessing Time Travel Narratives for Environmental Sustainability Education. In A. E. J. Wals & P. B. Corcoran (Eds.), *Learning for Sustainability in Times of Accelerating Change* (pp. 91–104). Wageningen: Wageningen Academic Publishers.

Packard, V. (1957). *The Hidden Persuaders*. New York: D. McKay Co.

Panofsky, E. (1927/1997). *Perspective as Symbolic Form*. New York: Zone Books.

Parr, A. (2009). *Hijacking Sustainability*. Cambridge, MA: MIT Press.

Pasquale, F. (2015). *The Black Box Society: The Secret Algorithms That Control Money and Information*. Cambridge, MA: Harvard University Press.

Patterson, R., Winterbottom, M. D., & Pierce, B. J. (2006). Perceptual Issues in the Use of Head-Mounted Visual Displays. *Human Factors: The Journal of the Human Factors and Ergonomics Society, 48*(3), 555–573.

Pearce, W., Holmberg, K., Hellsten, I., & Nerlich, B. (2014). Climate Change on Twitter: Topics, Communities and Conversations About the 2013 IPCC Working Group 1 Report. *PLoS One, 9*(4), e94785.

Peffer, T., Pritoni, M., Meier, A., Aragon, C., & Perry, D. (2011). How People Use Thermostats in Homes: A Review. *Building and Environment, 46*(12), 2529–2541.

Pettersen, I. N., & Boks, C. (2008). The Ethics in Balancing Control and Freedom When Engineering Solutions for Sustainable Behaviour. *International Journal of Sustainable Engineering, 1*(4), 287–297.

Pierce, J., Odom, W., & Blevis, E. (2008). Energy Aware Dwelling: A Critical Survey of Interaction Design for Eco-visualizations. Paper Presented at OZCHI 2008, Cairns.

Pinch, T. J., & Bijker, W. E. (1984). The Social Construction of Facts and Artefacts: Or How the Sociology of Science and the Sociology of Technology Might Benefit Each Other. *Social Studies of Science, 14*, 399–441.

Pogue, D. (2012, January 18). Use It Better: The Worst Tech Predictions of All Time. *Scientific American*. Retrieved from https://www.scientificamerican.com/article/pogue-all-time-worst-tech-predictions

Pollan, M. (2008, April 20). Why Bother? *New York Times*. Retrieved from http://www.nytimes.com/2008/04/20/magazine/20wwln-lede-t.html?_r=3&

Prost, S., Schrammel, J., & Tscheligi, M. (2014). 'Sometimes It's the Weather's Fault': Sustainable HCI & Political Activism, CHI '14 Extended Abstracts on Human Factors in Computing Systems (pp. 2005–2010). New York: ACM.

Redström, J. (2008). RE: Definitions of Use. *Design Issues, 29*(4), 410–423.

Rees, W. E. (2012). Cities as Dissipative Structures: Global Change and the Vulnerability of Urban Civilization. In M. P. Weinstein & R. E. Turner (Eds.), *Sustainability Science: The Emerging Paradigm and the Urban Environment* (pp. 247–273). New York: Springer.

Regan, T. (2016, November 22). Eagle Flight. *Trusted Reviews*. Retrieved from http://www.trustedreviews.com/reviews/eagle-flight

Reibelt, L. M., Richter, T., Rendigs, A., & Mantilla-Contreras, J. (2017). Malagasy Conservationists and Environmental Educators: Life Paths into Conservation. *Sustainability, 9*(2), article #227.

Relph, E. (2008). *Place and Placelessness* (Reprinted ed.). London: Pion Limited.

Renn, O. (2011). The Social Amplification/Attenuation of Risk Framework: Application to Climate Change. *Wiley Interdisciplinary Reviews: Climate Change, 2*(2), 154–169.

Reser, J. P., Morrissey, S. A., & Ellul, M. (2011). The Threat of Climate Change: Psychological Response, Adaptation, and Impacts. In I. Weissbecker (Ed.), *Climate Change and Human Well-Being: Global Challenges and Opportunities* (pp. 19–42). New York/Dordrecht/Heidelberg/London: Springer.

Reser, J. P., Bradley, G. L., & Ellul, M. C. (2014). Encountering Climate Change: 'Seeing' Is More Than 'Believing'. *Wiley Interdisciplinary Reviews: Climate Change, 5*(4), 521–537.

Rheingold, H. (1991). *Virtual Reality*. New York: Summit Books.

Riley, T. (2017, July 10). Just 100 Companies Responsible for 71% of Global Emissions, Study Says. *The Guardian*. Retrieved from https://www.theguardian.com/sustainable-business/2017/jul/10/100-fossil-fuel-companies-investors-responsible-71-global-emissions-cdp-study-climate-change

Rittel, H. W. J., & Webber, M. (1973). Dilemmas in a General Theory of Planning. *Policy Sciences, 4*, 155–169.

Robertson, A. (2017, January 26). The Best Virtual Reality from the 2017 Sundance Film Festival. *The Verge*. Retrieved from https://www.theverge.com/2017/1/26/14396976/best-vr-sundance-film-festival-2017

Robinson, J. (2004). Squaring the Circle? Some Thoughts on the Idea of Sustainable Development. *Ecological Economics, 48*, 369–384.

Robinson, J. (2008). Being Undisciplined: Transgressions and Intersections in Academia and Beyond. *Futures, 40*(1), 70–86.

Robinson, J., & Cole, R. J. (2015). Theoretical Underpinnings of Regenerative Sustainability. *Building Research & Information, 43*(2), 133–143.

Robinson, J., Carmichael, J., VanWynsberghe, R., Tansey, J., Journeay, M., & Rogers, L. (2006). Sustainability as a Problem of Design: Interactive Science in the Georgia Basin. *The Integrated Assessment Journal, 6*(4), 165–192.

Robinson, J., Burch, S., Talwar, S., O'Shea, M., & Walsh, M. (2011). Envisioning Sustainability: Recent Progress in the Use of Participatory Backcasting Approaches for Sustainability Research. *Technological Forecasting and Social Change, 78*(5), 756–768.

Rockström, J. (2015, November 14). The Planet's Future Is in the Balance. But a Transformation Is Already Under Way. *The Guardian*. Retrieved from https://www.theguardian.com/environment/2015/nov/14/un-climate-change-summit-paris-planet-future-balance-science

Rockström, J., & Klum, M. (2015). *Big World, Small Planet: Abundance Within Planetary Boundaries*. New Haven: Yale University Press.

Rockström, J., Steffen, W., Noone, K., Persson, Å., Chapin, F. S., III, Lambin, E. F., et al. (2009). A Safe Operating Space for Humanity. *Nature, 461*, 472–475.

Rorty, R. (2007). Philosophy as a Transitional Genre. In *Philosophy as Cultural Politics* (pp. 3–28). Cambridge: Cambridge University Press.

Rothman, D. S., Robinson, J., & Biggs, D. (2002). Signs of Life: Linking Indicators and Models in the Context of QUEST. In H. Abaza & A. Baranzini (Eds.), *Implementing Sustainable Development, Integrated Assessment and Participatory Decision-Making Processes* (pp. 182–199). Cheltenham: Edward Elgar.

Rowsome, A. (2017, May 10). Can Virtual Reality Help Us Tackle Climate Change? *Vice Impact*. Retrieved from https://impact.vice.com/en_us/article/xyeg97/can-virtual-reality-help-us-tackle-climate-change

Rozendaal, M. (2016). Objects with Intent: A New Paradigm for Interaction Design. *Interactions, 23*(3), 62–65.

Ruijten, P. A. M., Midden, C. J. H., & Ham, J. (2011). Unconscious Persuasion Needs Goal-Striving: The Effect of Goal Activation on the Persuasive Power of Subliminal Feedback. In *Proceedings of Persuasive 2011 (article number 4)*. New York: ACM.

Rutherford, A. (2003). B. F. Skinner's Technology of Behavior in American Life: From Consumer Culture to Counterculture. *Journal of History of the Behavioral Sciences, 39*(1), 1–23.

Samuel, L. R. (2010). *Freud on Madison Avenue: Motivation Research and Subliminal Advertising in America*. Philadelphia: University of Pennsylvania Press.

Sapieha, C. (2016, November 9). Eagle Flight Review: Time Keeps on Slipping in Ubisoft's Repetitive VR Simulation of Bird Life. *Financial Post*. Retrieved from http://business.financialpost.com/technology/gaming/eagle-flight-review-time-keeps-on-slipping-in-ubisofts-repetitive-vr-simulation-of-bird-life

Sartre, J.-P. (2004). *The Imaginary: A Phenomenological Psychology of the Imagination* (trans: Webber, J.). London/New York: Routledge.

Sassoon, D. (2012, August 20). Crude, Dirty and Dangerous. *The New York Times*. Retrieved from http://www.nytimes.com/2012/08/21/opinion/the-dangers-of-diluted-bitumen-oil.html

Schäfer, M. S. (2012). Online Communication on Climate Change and Climate Politics: A Literature Review. *Wiley Interdisciplinary Reviews: Climate Change*, *3*(6), 527–543.

Schroeder, R. (1996). *Possible Worlds: The Social Dynamic of Virtual Reality Technology*. Bolder: Westview Press.

Schueneman, T. (2013, July 26). Future of San Francisco's Market Street Comes into View. *Triple Pundit*. Retrieved from http://www.triplepundit.com/2013/07/future-san-franciscos-market-street-comes-view/

Schuldt, J. P., & Roh, S. (2014). Of Accessibility and Applicability: How Heat-Related Cues Affect Belief in "Global Warming" Versus "Climate Change". *Social Cognition, 32*(3), 217–238.

Schuldt, J. P., Konrath, S. H., & Schwarz, N. (2011). "Global Warming" or "Climate Change"?: Whether the Planet Is Warming Depends on Question Wording. *Public Opinion Quarterly, 75*(1), 115–124.

Schüll, N. D. (2013). *Addiction by Design: Machine Gambling in Las Vegas*. Princeton/Oxford: Princeton University Press.

Schwab, K. (2017). *The Fourth Industrial Revolution*. New York: Crown Business.

Scott, J. C. (1998). *Seeing Like a State: How Certain Schemes to Improve the Human Condition Have Failed*. New Haven/London: Yale University Press.

Scott, A. C. (2007). *The Nonlinear Universe: Chaos, Emergence, Life*. Berlin/Heidelberg: Springer.

Seamon, D. (1979). *A Geography of the Lifeworld: Movement, Rest and Encounter*. London: Croom Helm.

Seamon, D. (2015). Situated Cognition and the Phenomenology of Place: Lifeworld, Environmental Embodiment, and Immersion-in-World. *Cognitive Processing, 16*(supplement 1), 389–392.

Senecah, S. (2004). The Trinity of Voice: The Role of Practical Theory in Planning and Evaluating the Effectiveness of Environmental Participatory Processes. In S. P. Depoe, J. W. Delicath, & M.-F. A. Elsenbeer (Eds.), *Communication and Public Participation in Environmental Decision Making* (pp. 13–33). Albany: State University of New York Press.

Shapka, J. D., Law, D. M., & VanWynsberghe, R. (2008). Quest for Communicating Sustainability: Gb-Quest as a Learning Tool for Effecting Conceptual Change. *Local Environment, 13*(2), 107–127.

Sheppard, S. R. J. (2001). Guidance for Crystal Ball Gazers: Developing a Code of Ethics for Landscape Visualization. *Landscape and Urban Planning, 54*(1), 183–199.

Sheppard, S. R. J. (2005). Landscape Visualisation and Climate Change: The Potential for Influencing Perceptions and Behaviour. *Environmental Science and Policy, 8,* 637–654.

Shove, E. (2010a). Beyond the ABC: Climate Change Policy and Theories of Social Change. *Environment and Planning A, 42,* 1273–1285.

Shove, E. (2010b). Social Theory and Climate Change: Questions Often, Sometimes and Not Yet Asked. *Theory, Culture & Society, 27*(2–3), 277–288.

Shove, E. (2011). On the Difference Between Chalk and Cheese – A Response to Whitmarsh et al.'s Comments on 'Beyond the ABC: Climate Change Policy and Theories of Social Change'. *Environment and Planning A, 43,* 262–264.

Shove, E., & Spurling, N. (Eds.). (2013). *Sustainable Practices: Social Theory and Climate Change.* London/New York: Routledge.

Silberman, M. S., Blevis, E., Huang, E., Nardi, B. A., Nathan, L. P., Busse, D., et al. (2014a). What Have We Learned? A SIGCHI HCI & Sustainability Community Workshop. *CHI '14 Extended Abstracts on Human Factors in Computing Systems* (pp. 143–146). New York: ACM.

Silberman, M. S., Knowles, B., Nathan, L., Bendor, R., Clear, A., Hakansson, M., et al. (2014b). Next Steps for Sustainable HCI. *Interactions, 21*(5), 66–69.

Sisson, P. (2015, December 14). Check Out Buckminster Fuller's Simulation to Save the Planet. *Curbed.* Retrieved from http://www.curbed. com/2015/12/14/10621262/buckminster-fuller-world-games-columbia

Skarda, E. (2011, October 21). Top 10 Failed Predictions: Technology? What's That? *Time Magazine.* Retrieved from http://content.time.com/time/specials/packages/article/0,28804,2097462_2097456_2097467,00.html

Skinner, B. F. (1971). *Beyond Freedom and Dignity.* Indianapolis/Cambridge: Hackett.

Sloman, A. (2002). The Irrelevance of Turing Machines to Artificial Intelligence. In M. Scheutz (Ed.), *Computationalism: New Directions* (pp. 87–127). Cambridge, MA: MIT Press.

Slovic, P. (1987). Perception of Risk. *Science, 236*(4799), 280–285.

Smids, J. (2012). The Voluntariness of Persuasive Technology. In M. Bang & E. L. Ragnemalm (Eds.), *PERSUASIVE 2012* (pp. 123–132). Heidelberg: Springer.

Snow, S., Buys, L., Roe, P., & Brereton, M. (2013). Curiosity to Cupboard: Self Reported Disengagement with Energy Use Feedback Over Time. In *Proceedings of OzCHI '13* (pp. 245–254). New York: ACM.

Solnit, R. (2016). *Hope in the Dark: Untold Histories, Wild Possibilities* (3rd ed.). Chicago: Haymarket Books.

Spartz, J. T., Su, L. Y.-F., Griffin, R., Brossard, D., & Dunwoody, S. (2017). YouTube, Social Norms and Perceived Salience of Climate Change in the American Mind. *Environmental Communication, 11*(1), 1–16.

Stagoll, C. (2005). Concepts. In A. Parr (Ed.), *The Deleuze Dictionary* (pp. 50–51). Edinburgh: Edinburgh University Press.

Starosielski, N., & Walker, J. (Eds.). (2016). *Sustainable Media: Critical Approaches to Media and Environment.* New York: Routledge.

Stephens, S. H., DeLorme, D. E., & Hagen, S. C. (2017). Evaluation of the Design Features of Interactive Sea-Level Rise Viewers for Risk Communication. *Environmental Communication, 11*(2), 248–262.

Stern, P. C. (2000). Toward a Coherent Theory of Environmentally Significant Behavior. *Journal of Social Issues, 56*(3), 407–424.

Stevenson, K. T., Peterson, M. N., Carrier, S. J., Strnad, R. L., Bondell, H. D., Kirby-Hathaway, T., & Moore, S. E. (2014). Role of Significant Life Experiences in Building Environmental Knowledge and Behavior Among Middle School Students. *The Journal of Environmental Education, 45*(3), 163–177.

Stiegler, B. (1998). *Technics and Time (Vol. 1: The Fault of Epimetheus)* (trans: Beardsworth, R.). Stanford: Stanford University Press.

Stinson, J. (2017). Re-creating Wilderness 2.0: Or Getting Back to Work in a Virtual Nature. *Geoforum, 79,* 174–187.

Stirling, A. (2006). Analysis, Participation and Power: Justification and Closure in Participatory Multi-Criteria Analysis. *Land Use Policy, 23,* 95–107.

Stolterman, E. (2008). The Nature of Design Practice and Implications for Interaction Design Research. *International Journal of Design, 2*(1), 55–65.

Strengers, Y. (2011). Designing Eco-feedback Systems for Everyday Life. In *Proceedings of CHI 2011* (pp. 2135–2144). New York: ACM.

Strengers, Y., & Maller, C. (Eds.). (2015). *Social Practices, Intervention and Sustainability: Beyond Behaviour Change.* London/New York: Routledge.

Strike, K. A. (1975). Beyond Freedom and Dignity. *Studies in Philosophy and Education, 9*(1–2), 112–137.

Suzuki, D. T. (2007). *The Sacred Balance: Rediscovering Our Place in Nature* (3rd ed.). Vancouver: David Suzuki Foundation/Greystone Books.

Swart, R. J., Raskin, P., & Robinson, J. (2004). The Problem of the Future: Sustainability Science and Scenario Analysis. *Global Environmental Change, 14,* 137–146.

Talwar, S., Wiek, A., & Robinson, J. (2011). User Engagement in Sustainability Research. *Science and Public Policy, 38*(5), 379–390.

Tanenbaum, J. (2014). Design Fictional Interactions: Why HCI Should Care About Stories. *Interactions, 21*(5), 22–23.

Tavris, C., & Aronson, E. (2007). *Mistakes Were Made (But Not by Me): Why We Justify Foolish Beliefs, Bad Decisions, and Hurtful Acts.* Orlando: Harcourt.

Taylor, C. (2004). *Modern Social Imaginaries.* Durham: Duke University Press.

Taylor, P. J. (2005). *Unruly Complexity: Ecology, Interpretation, Engagement.* Chicago: University of Chicago Press.

Terzano, K., & Morckel, V. (2016). SimCity in the Community Planning Classroom: Effects on Student Knowledge, Interests, and Perceptions of the Discipline of Planning. *Journal of Planning Education and Research, 37*(1), 95–105.

Thaler, R. H. (2000). From Homo Economicus to Homo Sapiens. *Journal of Economic Perspectives, 14*(1), 133–141.

Thaler, R. H., & Sunstein, C. R. (2008). *Nudge: Improving Decisions About Health, Wealth, and Happiness.* New Haven: Yale University Press.

Thiel, P. (1996). *People, Paths, and Purposes: Notations for a Participatory Envirotecture.* Seattle/London: University of Washington Press.

Thiele, L. P. (2016). *Sustainability* (2nd ed.). Cambridge/Malden: Polity Press.

Thieme, A., Comber, R., Miebach, J., Weeden, J., Krämer, N., Lawson, S., & Olivier, P. (2012). "We've Bin Watching You": Designing for Reflection and Social Persuasion to Promote Sustainable lifestyles. In *Proceedings of CHI 2012* (pp. 2337–2346). New York: ACM.

Timmer, J., Kool, L., & van Est, R. (2015). Ethical Challenges in Emerging Applications of Persuasive Technology. In T. MacTavish & S. Basapur (Eds.), *Persuasive Technology. PERSUASIVE 2015* (pp. 196–201). Cham: Springer.

Tomlinson, B. (2010). *Greening Through IT: Information Technology for Environmental Sustainability.* Cambridge, MA: MIT Press.

Tresch, J. (2007). Technological World-Pictures: Cosmic Things and Cosmograms. *Isis, 98*(1), 84–99.

Trischler, H. (2016). The Anthropocene: A Challenge for the History of Science, Technology, and the Environment. *NTM, 24*(24), 309–335.

Tromp, N., Hekkert, P., & Verbeek, P.-P. (2011). Design for Socially Responsible Behavior: A Classification of Influence Based on Intended User Experience. *Design Issues, 27*(3), 3–19.

Tuan, Y.-f. (1977). *Space and Place: The Perspective of Experience.* Minneapolis: University of Minnesota Press.

Turner, C. (2007). *The Geography of Hope: A Tour of the World We Need.* Toronto: Random House Canada.

Tversky, A., & Kahneman, D. (1982). Judgment Under Uncertainty: Heuristics and Biases. In D. Kahneman, P. Slovic, & A. Tversky (Eds.), *Judgment Under Uncertainty: Heuristics and Biases* (pp. 3–20). Cambridge/New York: Cambridge University Press.

UN. (2015). *Transforming Our World: The 2030 Agenda for Sustainable Development.* Retrieved from https://sustainabledevelopment.un.org/content/documents/21252030%20Agenda%20for%20Sustainable%20Development%20web.pdf

UNWCED. (1987). *Our Common Future: Report of the World Commission on Environment and Development.* New York: Oxford University Press.

Uricchio, W., Wolozin, S., Bui, L., Flynn, S., & Tortum, D. (2015). *Mapping the Intersection of Two Cultures: Interactive Documentary and Digital Journalism.* Retrieved from http://opendoclab.mit.edu/interactivejournalism/

Vaillant, J. (2005). *The Golden Spruce: A True Story of Myth, Madness and Greed.* New York: W.W. Norton.

van Kerkhoff, L., & Lebel, L. (2006). Linking Knowledge and Action for Sustainable Development. *Annual Review of Environment and Resources, 31,* 445–477.

Vancouver. (2010). *Vancouver 2020: A Bright Green Future.* Vancouver: City of Vancouver.

VanWynsberghe, R., Carmichal, J., & Khan, S. (2007). Conceptualizing Sustainability: Simulating Concrete Possibilities in an Imperfect World. *Local Environment, 12*(3), 279–293.

Vaughn, R. (1980). How Advertising Works: A planning model. *Journal of Advertising Research, 20*(5), 27–33.

Verbeek, P.-P. (2005). *What Things Do: Philosophical Reflections on Technology, Agency, and Design* (trans: Crease, R. P.). University Park: Pennsylvania State University Press.

Verbeek, P.-P. (2006). Persuasive Technology and Moral Responsibility: Toward an Ethical Framework for Persuasive Technologies. Paper Presented at Persuasive 06, Eindhoven, The Netherlands.

Verbeek, P.-P. (2015). Beyond Interaction: A Short Introduction to Mediation Theory. *Interactions, 12*(3), 26–31.

Vervoort, J. M., Bendor, R., Kelliher, A., Strik, O., & Helfgott, A. E. R. (2015). Scenarios and the Art of Worldmaking. *Futures, 74,* 62–70.

Wagner, G. (2011, September 7). Going Green But Getting Nowhere. *New York Times.* Retrieved from http://www.nytimes.com/2011/09/08/opinion/going-green-but-getting-nowhere.html

Walker, G. B. (2007). Public Participation as Participatory Communication in Environmental Policy Decision-Making: From Concepts to Structured Conversations. *Environmental Communication, 1*(1), 99–110.

Wallace-Wells, D. (2017, July 9). The Uninhabitable Earth. *New York Magazine.* Retrieved from http://nymag.com/daily/intelligencer/2017/07/climate-change-earth-too-hot-for-humans.html

Wals, A. E. J., & Corcoran, P. B. (2012). Re-orienting, Re-connecting and Re-imagining: Learning-Based Responses to the Challenge of (Un)sustainability. In A. E. J. Wals & P. B. Corcoran (Eds.), *Learning for Sustainability in Times of Accelerating Change* (pp. 21–32). Wageningen: Wageningen Academic Publishers.

Weber, E. U. (2006). Experience-Based and Description-Based Perceptions of Long-Term Risk: Why Global Warming Does Not Scare Us (Yet). *Climatic Change, 77*(1–2), 103–120.

Weber, E. U. (2010). What Shapes Perceptions of Climate Change? *Wiley Interdisciplinary Reviews: Climate Change, 1*(3), 332–342.

Weber, E. U. (2016). What Shapes Perceptions of Climate Change? New Research Since 2010. *Wiley Interdisciplinary Reviews: Climate Change, 7*(1), 125–134. https://doi.org/10.1002/wcc.377.

Weisman, A. (2007). *The World Without Us.* New York: Picador.

Weizenbaum, J. (1976). *Computer Power and Human Reason: From Judgment to Calculation.* San Francisco: W. H. Freeman.

Wells, M. (2016). Deliberate Constructions of the Mind: Simulation Games as Fictional Models. *Games and Culture, 11*(5), 528–547.

Wells, N. M., & Lekies, K. S. (2006). Nature and the Life Course: Pathways from Childhood Nature Experiences to Adult Environmentalism. *Children, Youth and Environments, 16*(1), 1–24.

Westley, F., Carpenter, S. R., Brock, W. A., Holling, C. S., & Gunderson, L. H. (2002). Why Systems of People and Nature Are Not Just Social and Ecological Systems. In L. H. Gunderson & C. S. Holling (Eds.), *Panarchy: Understanding Transformations in Human and Natural Systems.* Washington, DC: Island Press.

Whitmarsh, L. (2008). Are Flood Victims More Concerned About Climate Change Than Other People? The Role of Direct Experience in Risk Perception and Behavioural Response. *Journal of Risk Research, 11*(3), 351–374.

Whitmarsh, L. (2009). What's in a Name? Commonalities and Differences in Public Understanding of "Climate Change" and "Global Warming". *Public Understanding of Science, 18*(4), 401–420.

Whitmarsh, L., O'Neill, S., & Lorenzoni, I. (2011). Climate Change or Social Change? Debate Within, Amongst, and Beyond Disciplines. *Environment and Planning A, 43*, 258–261.

Williams, C. C., & Chawla, L. (2016). Environmental Identity Formation in Nonformal Environmental Education Programs. *Environmental Education Research, 22*(7), 978–1001.

Williams, A., & Srnicek, N. (2013). # Accelerate Manifesto for an Accelerationist Politics. *Critical Legal Thinking, 14*, 72–98.

Wilson, G. T., Lilley, D., & Bhamra, T. A. (2013). Design Feedback Interventions for Household Energy Consumption Reduction. Paper Presented at the ERSCP-EMSU 2013 Conference, Istanbul.

Wilson, G. T., Bhamra, T., & Lilley, D. (2015). The Considerations and Limitations of Feedback as a Strategy for Behaviour Change. *International Journal of Sustainable Engineering, 8*(3), 186–195.

Winograd, T., & Flores, F. (1986). *Understanding Computers and Cognition: A New Foundation for Design.* Norwood: Ablex Pub. Corp.

Wittgenstein, L. (2001). *Philosophical Investigations* (trans: Anscombe, G. E. M., 3rd ed.). Oxford/Malden: Blackwell.

Woessner, M. (2015). Teaching with SimCity: Using Sophisticated Gaming Simulations to Teach Concepts in Introductory American Government. *PS: Political Science & Politics, 48*(2), 358–363.

Wohlberg, M. (2013, December 2). Ideas Battle Online for Control of Fort McMoney. *Northern Journal.* Retrieved from https://norj.ca/2013/12/ideas-battle-online-for-control-of-fort-mcmoney/

Worster, D. (1994). *Nature's Economy: A History of Ecological Ideas* (2nd ed.). Cambridge/New York: Cambridge University Press.

Wright, C., Nyberg, D., De Cock, C., & Whiteman, G. (2013). Future Imaginings: Organizing in Response to Climate Change. *Organization, 20*(5), 647–658.

Wynne, B. (1992). Misunderstood Misunderstanding: Social Identities and Public Uptake of Science. *Public Understanding of Science, 1*(3), 281–304.

Yang, R., Newman, M. W., & Forlizzi, J. (2014). Making Sustainability Sustainable: Challenges in the Design of Eco-interaction Technologies. In *Proceedings of CHI 2014* (pp. 823–832). New York: ACM.

Yusoff, K., & Gabrys, J. (2011). Climate Change and the Imagination. *Wiley Interdisciplinary Reviews: Climate Change, 2*, 516–534.

Zachrisson, J., & Boks, C. (2012). Exploring Behavioural Psychology to Support Design for Sustainable Behaviour Research. *Journal of Design Research, 10*(1–2), 50–66.

Zapico, J. L., Turpeinen, M., & Brandt, N. (2009). Climate Persuasive Devices: Changing Behaviour Towards Low-Carbon Lifestyles. In S. Chatterjee & P. Dev (Eds.), *Proceedings of Persuasive '09* (Article 14). New York: ACM Press.

Zappen, J. P. (2005). Digital Rhetoric: Toward an Integrated Theory. *Technical Communication Quarterly, 14*(3), 319–325.

Index[1]

[1] Note: Page numbers followed by 'n' refer to footnotes.

© The Author(s) 2018   215
R. Bendor, *Interactive Media for Sustainability*,
Palgrave Studies in Media and Environmental Communication,
https://doi.org/10.1007/978-3-319-70383-1